MW01260010

Hummingbirds

A Celebration of Nature's Jewels

Glenn Bartley and Andy Swash

Contributors: Rob Hume, Christopher J. Sharpe and Robert Still
Illustrations by Jeanne Melchels

WILDGuides

PRINCETON
press.princeton.edu

Published by Princeton University Press,
41 William Street, Princeton, New Jersey 08540
99 Banbury Road, Oxford OX2 6JX
press.princeton.edu

British Library Cataloging-in-Publication Data is available

Library of Congress Control Number 2022930986
ISBN 978-0-691-18212-4
ISBN (ebook) 978-0-691-22560-9

Production and design by WILD*Guides* Ltd., Old Basing, Hampshire UK
Printed in Italy

10 9 8 7 6 5 4 3 2 1

FRONT COVER: **Black-billed Streamertail** *Trochilus scitulus* | Jamaica
BACK COVER: **Rufous-crested Coquette** *Lophornis delattrei* | Peru
FRONTISPIECE: **Fiery-throated Hummingbird** *Panterpe insignis* | Costa Rica

Contents

Foreword 5

Preface 7

An Introduction to Hummingbirds 9

Adaptations for Exceptional Lifestyles 15

Color and Iridescence: Flashes of Brilliance 49

Breeding: Continuing the Line 87

Biogeography and Biodiversity: The Hummingbirds' Realm 115

Hummingbirds and People: History, Discovery and Culture 179

Conservation: Hummingbirds Under Threat 191

In Pursuit of Hummingbirds: A Photographic Journey—Glenn Bartley 211

In Search of Nature's Jewels: A Personal Quest—Andy Swash 229

Taxonomy: The BirdLife List of Species 239

Appendix 1: Misnamed Hummingbirds 274

Appendix 2: Named Hummingbird Hybrids 275

Further Reading and Sources of Useful Information 277

Acknowledgements and Photo Credits 279

Index 283

BirdLife INTERNATIONAL The names of the hummingbirds and the taxonomy used throughout this book follow BirdLife International (see *page 239*).

All the images shown are males unless captioned otherwise.

The location given for each photograph indicates where it was taken. For information on the distribution of a particular species, see the relevant entry in *Taxonomy: The BirdLife List of Species* (*page 239*).

Foreword

Hummingbirds are adored by people around the world for their remarkable beauty, their often minute size, and their incredible flying capabilities. But while much is said of their beauty, in truth the services that hummingbirds perform for their ecosystems are far more valuable than the jewels after which many are named. Feasting on nectar, they play a vital role in pollinating plants, especially in cooler, high-altitude areas where insect pollinators tend to be far less numerous. This symbiotic relationship is so closely entwined that many flowers have evolved specifically to attract hummingbirds instead of insects. As a result, flowers are often brightly colored but unscented, since birds favor sight over smell. The relationship between these tiny birds and flowering plants benefits humans directly. In fact, approximately 5% of the plants we use for food or medicine worldwide are pollinated by birds, and this in turn plays a vital role in maintaining the wide range of habitats upon which we are so reliant. All of this means that, were we to lose a hummingbird, we would lose far more than just its beauty.

Yet, tragically, 39 species of hummingbird are threatened with extinction, nine of which are Critically Endangered, and they occur in a wide variety of habitats ranging from the dwindling forests of the tropical Andes to remote Pacific islands. If this was not sufficient cause for concern, the crisis is continuing to spread. Only two years ago, the Rufous Hummingbird, a species once common across North America, was reclassified as Near Threatened owing to a shocking decline in its population. As with many birds, intensive agriculture is decimating the insects on which this hummingbird relies during the breeding season. In addition, the impacts of climate change are beginning to be seen. For example, recent studies have shown that flowers are blooming as many as two weeks early in some locations. This means that, following their northward migration, many Rufous Hummingbirds arrive too late to take advantage of the peak in nectar production of this vital food source.

Habitat destruction is also putting pressures on already stressed populations of many hummingbirds. There is always hope, however, and I am immensely proud of the role that BirdLife Partners and other collaborators are playing in bringing threatened hummingbirds back from the brink of extinction. In the high Andean forests of my home country of Ecuador, where hummingbirds are an ever-present delight, BirdLife in Ecuador, Aves y Conservación, is employing local people to cultivate over 4,500 native plants to help in restoring the habitat of the Black-breasted Puffleg, one of the rarest birds in the world. BirdLife in Brazil, SAVE Brasil (Sociedade para a Conservação das Aves do Brasil) has been actively involved in restoring areas of degraded Atlantic Forest, with dramatic results: in one 40-hectare patch the number of bird species skyrocketed from three to seventy, including nine species of hummingbird that fertilize flowers, thereby helping to maintain a diverse habitat in a beneficial cycle. This just goes to show that conservation action based on solid science really does work.

As with so many of the birds in the world, there is still a great deal to be learned about the biology, ecology and habitat requirements of hummingbirds. Indeed, even in a country as well known for hummingbirds as is Ecuador, new species are still being found. I was very excited when I first heard of the discovery by Ecuadorian scientists of the stunningly beautiful Blue-throated Hillstar in a remote area of the high Andes. I soon realized, however, that owing to its restricted range it would likely already be under serious threat of extinction. It is therefore reassuring that the local conservation community, supported by BirdLife Partners such as American Bird Conservancy, was able to respond immediately in taking action to help to safeguard the habitat of this iconic bird.

A hummingbird's beauty is not necessarily its greatest power, but it does make the bird a captivating ambassador for nature, forging a strong connection with people of all backgrounds. The Black-breasted Puffleg was named the emblematic bird of Quito, Ecuador's capital, for this very reason. When we become inspired and spellbound by a hummingbird's splendor, we begin to think about its place in the natural world. It is becoming increasingly clear that we are entering a now-or-never era for climate and biodiversity action. My hope is that by reading this wonderful book, which draws extensively on BirdLife's science, and enjoying the fantastic photos, you will better understand how and where these amazing birds live and the challenges that they face. Please, then, consider what you can do personally, and together with others, to help to preserve them and our planet for future generations. It is, after all, safe to say that hummingbirds can help us, if we do what we can to help them.

Patricia Zurita
Chief Executive Officer
BirdLife International Partnership

FACING PAGE **Blue-throated Hillstar** *Oreotrochilus cyanolaemus* | Ecuador

Preface

> *"There is not, it may safely be asserted, in all the varied works of nature in her zoological productions, any family that can bear a comparison, for singularity of form, splendour of colour, or number and variety of species, with this the smallest of the feathered creation."*
>
> William Bullock, former curator of the London Natural History Museum, 1824

Hummingbirds are truly the jewels of the bird world, many having glittering iridescent feathers, the colors of which are reflected in their wonderful names—ruby, emerald, sapphire, topaz, amethyst and garnet. But they are much more than just beautiful birds: hummingbirds are incredible in so many ways, exhibiting a range of extreme adaptations to help them live their life in the fast lane.

The remarkable attributes of hummingbirds range from an extraordinarily high metabolic rate, exceeding that of any other animal, to skeletal adaptations that allow their unsurpassed powers of flight—notably the ability to fly in any direction, including backwards, and even upside down. Some can lower their metabolic rate and body temperature dramatically in order to survive periods of extreme cold. Hummingbirds are not, to be fair, among the world's best songsters—as they say, you can't have everything! Nevertheless, some do have energetic and spectacular courtship displays. Indeed, several smaller species perform high-speed dives, pulling huge g-forces that would tear other birds apart. Other hummingbirds undertake far-ranging migrations, including long-distance flights over the sea that seem barely credible for such tiny birds that require frequent inputs of energy just to stay alive. Add to this the tremendous diversity of shapes, forms and adornments among hummingbirds, and the rarity of many, some species having their entire world population restricted to just a tiny area, and it is not surprising that they are one of the most popular and sought-after groups of all birds—both with those who enjoy watching birds in their back yard and among hard-core birders.

It is their astonishing colors and unique behaviors and lifestyles that led to both of us being captivated by hummingbirds since an early age. Although we had previously corresponded regarding other book projects, our first meeting, back in June 2017, was entirely serendipitous. During a casual conversation while sipping caipirinhas and watching the mesmerizing antics of the many hummingbirds visiting the feeders at the delightful Eco-lodge Itororó in Brazil, we soon realized that between us we had the resources, skills and contacts to produce a book that really would do justice to these fantastic birds. Our concept was to provide a broad overview of the lives and fortunes of hummingbirds in a form that could be appreciated by birdwatchers and non-birdwatchers alike, showcasing a wide selection of the most spectacular photos ever taken.

But to ensure that the book was as up to date as possible, drawing on information available both in the published literature and from field ornithologists, and presented in an accessible style, we realized that we needed to draw on the skills and experience of others. We were therefore delighted when well-known writers Rob Hume and Chris Sharpe, top graphic designer Robert Still, the talented illustrator Jeanne Melchels and experienced editors Gill Swash and David Christie agreed to contribute to the project. All have made very significant and vital contributions to bringing the book to fruition.

Hummingbirds have for long been revered for their beauty, as conveyed so evocatively in the writings of the English naturalist and pioneer conservationist Charles Waterton, who traveled throughout Latin America in the early 19th century:

> *See it darting through the air almost as quick as thought!—now it is within a yard of your face!—in an instant gone!—now it flutters from flower to flower to sip the silver dew—it is now a ruby—now a topaz—now an emerald—now all burnished gold!*

Our hope is that, by reading this book and admiring the stunning photos, you will become as enthralled and enchanted by hummingbirds as Charles Waterton was nearly two hundred years ago, and as we both are today. The future of many of these fabulous birds is now in the balance and we should all consider carefully how we can help to ensure that they continue to thrive and play their part as nature's jewels in our wonderful world. If this book contributes, even if only in a small way, it will have been worth the effort.

Glenn Bartley and Andy Swash

FACING PAGE **Tufted Coquette** *Lophornis ornatus* | Trinidad

An Introduction to Hummingbirds

Hummingbirds are amazing creatures that are loved and admired by people worldwide. Their bright colors, glittering with iridescence, and their astonishingly aerobatic flight capabilities give them a unique appeal. Most hummingbirds are tiny, and included among them are the smallest birds in the world by some margin. The truth is that there is very little about a hummingbird that is ordinary and, as anyone who has observed one can attest, they certainly have charisma!

Hummingbirds are exclusively birds of the 'New World': found throughout the Americas, but nowhere else. In Africa, Asia and Australasia there are sunbirds, some of which look superficially similar to hummingbirds and to some extent fill the same ecological niche. Yet, as wonderful and colorful as most sunbirds are, they lack many of the physical adaptations that typify hummingbirds. Unlike hummingbirds, they are unable to hover for long periods and to fly backwards, the latter a trait exclusive to hummingbirds. Sunbirds are, in fact, much more closely related to, for example, pipits and finches; perhaps surprisingly, hummingbirds' closest relatives are the swifts and treeswifts.

Most people living in the Americas will know something about hummingbirds, and many will have had personal experience of them, since in most regions at least one species can be seen, the number increasing

Green-tailed Sunbird *Aethopyga nipalensis* | India

Golden-bellied Starfrontlet *Coeligena bonapartei* | Colombia

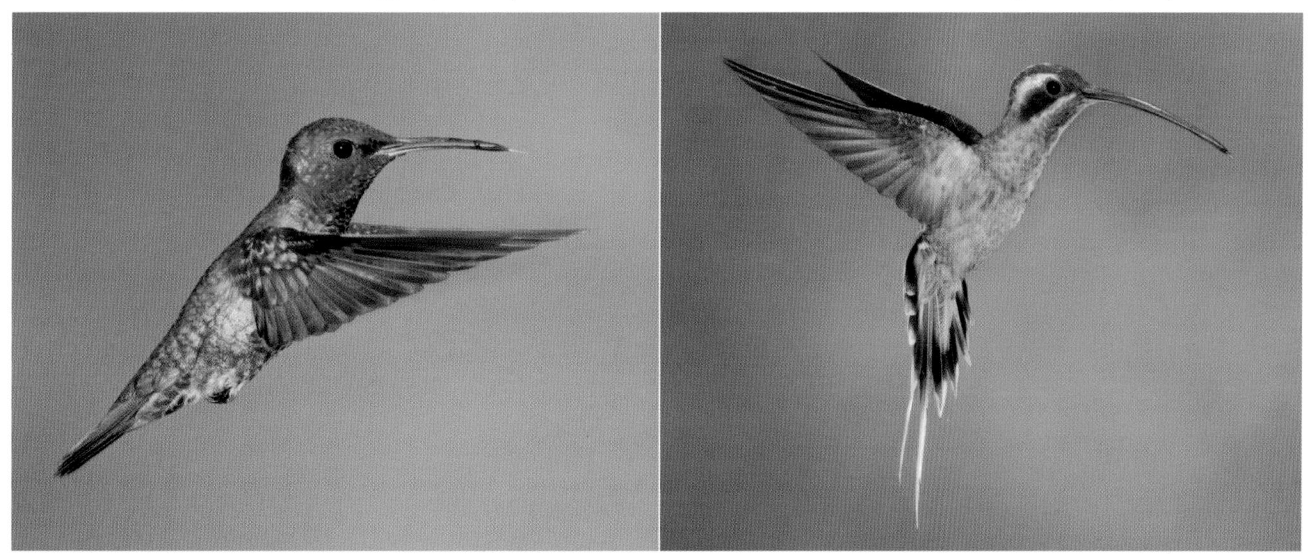

Golden-tailed Sapphire *Amazilia oenone* | Ecuador

Ecuadorian Hermit *Phaethornis baroni* | Ecuador

Although hummingbirds are distinctive in many ways, they do, at first glance, appear to have similar characteristics to those of a completely unrelated family of birds, the sunbirds (*top left*). Hummingbirds, however, are unique in their ability to hover for prolonged periods. Hummingbirds can be divided into two broad categories: the 'typical' hummingbirds (*top right* and *bottom left*), which tend to exhibit brightly colored, iridescent plumage; and the hermits (*bottom right*), which have a distinctive profile and are generally brownish or greenish.

FACING PAGE **Blue-chested Hummingbird** *Amazilia amabilis* | Panama

the closer you get to the equator. Indeed, in many areas hummingbirds are regular garden birds, even visiting urban yards. It is undoubtedly due to their familiarity, and in no small measure to their exquisite beauty, too, that hummingbirds have for long been revered—featuring in legends and myths, and in the traditions, ceremonies and rituals of many cultures for millennia. In more recent times, hummingbirds have continued to maintain their popularity and today are still strongly embedded in people's consciousness, being the subject of festivals, such as in the southern states of the USA, and also featuring frequently in pop-culture imagery. Even so, it seems still to be the case that in modern-day life hummingbirds are perhaps all too often taken for granted.

To people living elsewhere in the world, a hummingbird might simply be one of those exotic creatures of which they are vaguely aware. In recent decades, however, hummingbirds have become increasingly accessible, long-haul holidaymakers now having a chance of experiencing at least a few species firsthand. Trips to sunny destinations such as the Caribbean islands, or to the core hummingbird areas of Central and South America, take visitors into the heart of the hummingbirds' realm, and offer the incredible opportunity of seeing these fantastic birds, alive and free, whether in a hotel garden, a city green space, a specialist wildlife park where they are attracted to feeders, or out in the wild on a walk along a forest trail.

Birdwatchers, of course, increasingly target hummingbirds on their journeys in search of new species to see, to photograph or to study. There are growing numbers of conservation projects centered on hummingbirds in remote places that were little known to most naturalists just a few decades ago, and the organizations running these projects often welcome visitors. Despite this, and while much is being done to study the biology and ecology of hummingbirds, and to conserve the habitats where the most threatened species occur, there are still huge gaps in our knowledge.

But what exactly is a hummingbird? Early European explorers were puzzled when first confronted by creatures that seemed to be somehow intermediate between birds and insects. These strange and fascinating animals did, after all, shine with metallic colors, fly in all directions, and zip here and there at unimaginable speeds—attributes more normally associated with insects than with birds. They were sometimes seen feeding from flowers with what looked as much like a long proboscis as like a bird's bill, and the smallest species even looked remarkably like large moths. Modern-day naturalists can still be challenged by 'hummingbird mimics' such as sphinx moths in the genus *Aellopos*, which can easily be confused with hummingbirds, even by experienced observers.

Humming-bird and Humming-bird Hawk-moth.

The first person to give a scientific account of mimicry in animals was the English naturalist and explorer Henry Walter Bates. In his book *The Naturalist on the River Amazons*, published in 1864 (an illustration from which is shown *above*), he reports that it took him days to be confident of telling hummingbirds and hawkmoths apart, and that in the process he shot several hawkmoths by mistake!

Some hawkmoths, such as the **Titan Sphinx** *Aellopos titan* (*top*), which has a widespread distribution across the Americas, can look very similar to a small hummingbird, such as this female **Tufted Coquette** *Lophornis ornatus* (*bottom* | Trinidad)—and this has led to many birders spending much time puzzling over the identification of a 'mystery' hummingbird.

Of course, expedition naturalists quickly realized the truth: hummingbirds clearly had feathers, the fundamental characteristic of a bird.

The origin of hummingbirds

The origin of hummingbirds is the subject of much conjecture. The latest thinking is that they diverged from swifts and treeswifts, which on the basis of anatomical and morphological similarities appear to be their closest living relatives, around 42 million years ago, possibly in Eurasia.

The oldest known fossil hummingbirds, from 30–35 million years ago, have been found in Germany, Poland and France. Their fossil record, however, remains poor and it is possible that early hummingbirds occurred also outside Eurasia, although direct evidence for this has yet to be discovered.

Recent genetics-based research suggests that hummingbirds had firmly established their place in South America by around 22 million years ago. The fossil evidence of hummingbirds from this part of the world that is more than 10,000 years old is, however, extremely limited, and establishing their evolution in the Americas therefore remains problematic. On the basis of genetic research, hummingbirds appear to have pushed northwards into North America about 12 million years ago, this followed by several subsequent invasions along the Panamanian land-bridge, the Caribbean area being occupied only around five million years ago.

The plate tectonic activity that has led to the uplift and formation of the Andes and other ranges of mountains across the Americas over the past ten million years has also caused a progressive increase in the diversity of both habitats and climate, stratified by altitude as much as by latitude. In parallel with this, hummingbirds, and the flowers upon which they depend, continually coevolved, adapting to ever more varied opportunities, the early forms diverging into more and more species. One particularly remarkable feature of hummingbird ecology is that, across large parts of the Americas, an unusually large number of species is able to coexist within a relatively small area, having evolved differing lifestyles to exploit the many different microhabitats. Wherever the earliest species originated, the huge diversification of modern hummingbirds has clearly been a 'New World' phenomenon.

Although swifts, treeswifts and hummingbirds are all distantly related, the one feature that distinguishes hummingbirds, and probably mapped out their future, was that they evolved the ability to appreciate taste. Birds do not have the vertebrates' usual 'sweet taste receptor', but, and uniquely among birds, hummingbirds have an umami taste receptor (which in humans recognizes meaty or savory tastes) that is adapted to enable the detection of carbohydrates and sweetness. This taste for sugar,

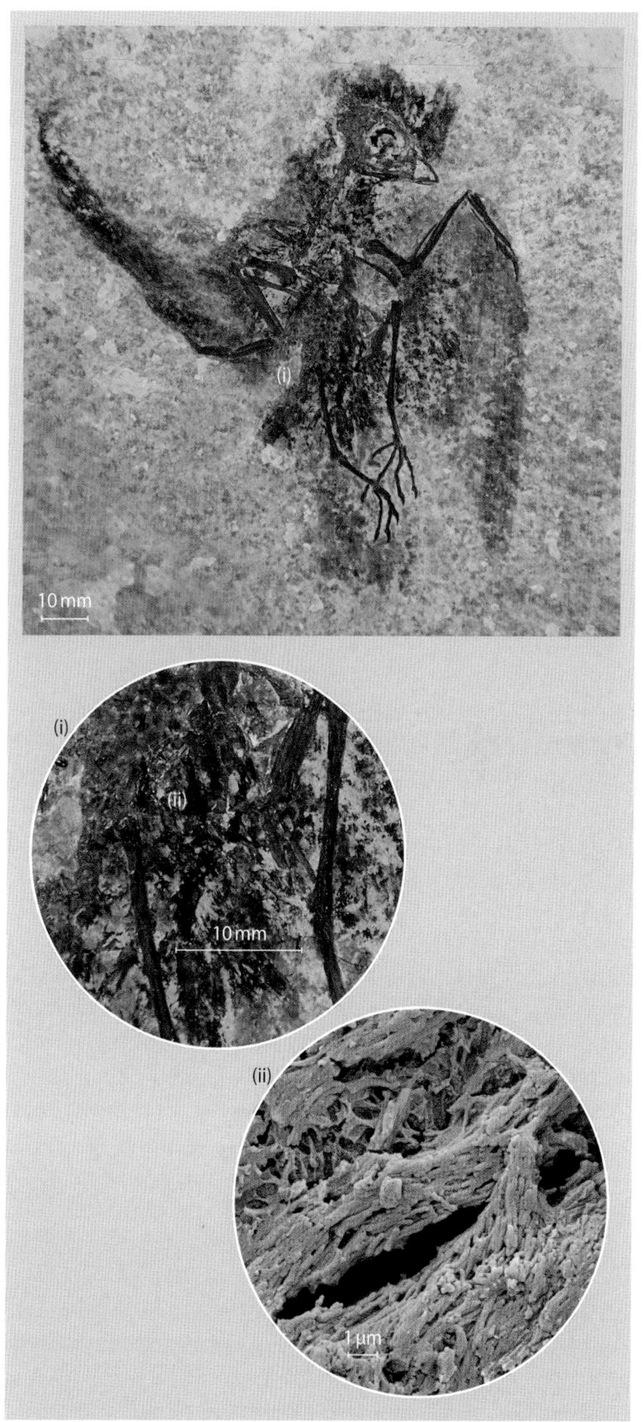

A fossil 50 million years old discovered in Wyoming, USA adds additional evidence to the origin of hummingbirds. It is apparently of a tiny bird that branched off the same evolutionary line as that which produced the treeswifts, swifts and hummingbirds (see *next page*). At less than 13 cm (5 inches) long and probably weighing no more than 28 g (1 oz), ***Eocypselus rowei*** evolved separately but left no modern-day descendants. Its wings appear to be intermediate between those of hummingbirds and the longer wings of swifts, indicating less specialized aerial behavior, and it clearly has an insect-eater's bill. The exceptionally preserved feathers exhibit detailed morphological structures, including rod-like eumelanosomes (shown in the lower Scanning Electron Microscope image) that are commonly associated with black, glossy black and some types of iridescent plumage in modern-day birds.

Images reproduced from a paper by Ksepka *et al.*, 2013 (see *page 277*), with the author's permission.

FACING PAGE A graphic summary of the current scientific knowledge of the evolutionary relationships between hummingbirds and their closest relatives, indicating when each order is believed to have diverged from a common ancestor [mya = million years ago]. NOTE: The figures given for the number of species do not include those that are extinct, and follows BirdLife International taxonomy as at December 2021.

hummingbirds' most insect-like attribute, meant that they became increasingly dependent upon a year-round source of nectar-rich flowers in order to survive (swifts and treeswifts feed only on flying insects and other airborne invertebrates). It may have taken many millions of years, but hummingbirds now constitute the second largest family of birds in the world (after the tyrant-flycatchers, Tyrannidae, another group of birds found only in the Americas), with 369 surviving species (currently, 450 species of tyrant-flycatcher are recognized).

Listing and describing: sorting the species

The study and classification of life forms on earth is called **taxonomy**. It is a science that involves describing living things as we know them, mapping out the evolutionary relationships (phylogeny), and assigning scientific names. The mapping process involves working backwards from existing species to construct a 'tree of life' that reflects their ancestry (a similar approach to the reconstruction of family trees in human genealogy). This process is, however, continually being refined as technology advances, as our knowledge increases, and as prevailing concepts change.

As a result of these studies, the animal **kingdom** is divided into groups, including the vertebrates (animals with backbones). The vertebrates are then subdivided into fishes, amphibians, reptiles, birds, and mammals. All birds are in the **class Aves**, which is subdivided into **orders**. Orders that are particularly closely related may be grouped together for convenience into **superorders**. Each order is further subdivided into **families** (some of which are divided further into **subfamilies**) and then **genera** (the plural of genus), and each genus is split into separate **species**—the category that is most familiar to us. Various taxonomic authorities recognize between 10,500 and 11,000 species of birds (although there are suggestions that the true figure may be much higher), in around 40 orders. In many cases, separate geographical populations of a species are distinguishable and these are described as **subspecies**. The term **clade**—a word derived from the Greek *klados*, which means a twig or branch—is used in phylogeny to denote a group of organisms that comprises *all* descendants from their most recent common ancestor.

In many cases the precise details of exactly what constitutes a species are still somewhat controversial. New methods of defining the relationships between species continue to emerge, particularly since the advent of genetic studies based on the analysis of DNA (deoxyribonucleic acid), the molecule that contains the hereditary material found in almost all living things. DNA has become a familiar everyday term, masking its complexity and the extraordinary avenues that it opens up for further research.

Hummingbird taxonomy is no exception to this controversy, and at both a species level, and even a genus level, it is still in a state of flux in light of the latest studies, and, of course, as the birds themselves continue to evolve and diversify. In fact, even the higher taxonomic groupings have been subject to rearrangement in recent years. Hummingbirds, swifts and treeswifts traditionally formed the order **Apodiformes** (which translates from Latin as *"footless"*) but it is now generally accepted that they are part of a larger, rather diverse group that includes nightjars: the superorder **Caprimulgimorphae**, as illustrated *opposite*.

Within the hummingbirds (family Trochilidae), six subfamilies are widely recognized. These are **Florisuginae**, the 'true' topazes and the jacobins, a group of 4 species; **Phaethornithinae**, the hermits, a distinctive group of 40 species; **Polytminae**, the mangos and relations, comprising 29 species including lancebills, violet-ears and fairies; **Lesbiinae**, the brilliants and coquettes, 129 species with a diverse range of characters from sunangels to sylphs and racket-tails to the astonishing Sword-billed Hummingbird; and **Trochilinae**, frequently referred to as trochilines, with 166 extant species. The subfamily Trochilinae is often further divided into three subgroups (or clades): **mountain-gems** (17 species), **bees** (37 species) and **emeralds** (112 species). The Giant Hummingbird does not fit into any of these groups and is therefore assigned its own subfamily, **Patagoninae**. Even though bird taxonomy is still evolving, with ever larger datasets being analyzed by means of increasingly sophisticated tools, it is predicted that this broad arrangement—or something very close to it—will stand the test of time. At any rate, for most observers these groupings provide a useful shorthand method of categorizing hummingbirds and gaining a basic understanding of their appearance and behavior.

A complete list of the extant species of hummingbird that are currently recognized by BirdLife International is included in the chapter *Taxonomy: The BirdLife List of Species* (*page 239*). It is important to recognize that this list reflects current knowledge as of December 2021 and that future changes are inevitable. This is not just because research is leading to different interpretations, but also because the hummingbirds themselves are changing. As they become ever more adapted to local environments, and especially to the plants within them, the tendency is for even greater diversification, initially into subspecies and eventually into new species. This is a continuing process and no-one really knows how much further it has to go.

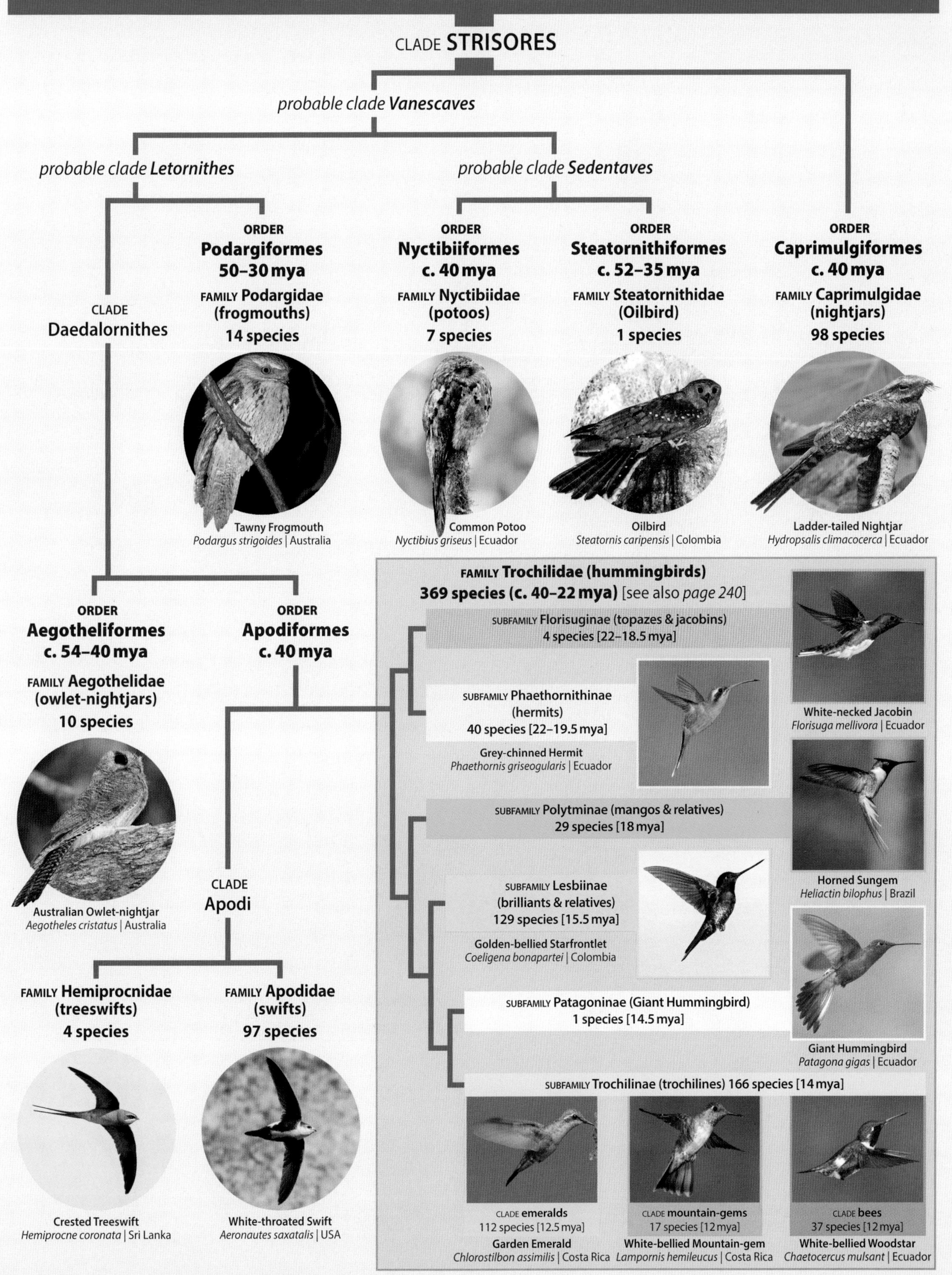

SUPERORDER CAPRIMULGIMORPHAE
CLADE STRISORES
probable clade Vanescaves
probable clade Letornithes
probable clade Sedentaves
ORDER Podargiformes 50–30 mya
FAMILY Podargidae (frogmouths)
14 species
ORDER Nyctibiiformes c. 40 mya
FAMILY Nyctibiidae (potoos)
7 species
ORDER Steatornithiformes c. 52–35 mya
FAMILY Steatornithidae (Oilbird)
1 species
ORDER Caprimulgiformes c. 40 mya
FAMILY Caprimulgidae (nightjars)
98 species
CLADE Daedalornithes
Tawny Frogmouth
Podargus strigoides | Australia
Common Potoo
Nyctibius griseus | Ecuador
Oilbird
Steatornis caripensis | Colombia
Ladder-tailed Nightjar
Hydropsalis climacocerca | Ecuador
ORDER Aegotheliformes c. 54–40 mya
FAMILY Aegothelidae (owlet-nightjars)
10 species
Australian Owlet-nightjar
Aegotheles cristatus | Australia
ORDER Apodiformes c. 40 mya
CLADE Apodi
FAMILY Hemiprocnidae (treeswifts)
4 species
FAMILY Apodidae (swifts)
97 species
Crested Treeswift
Hemiprocne coronata | Sri Lanka
White-throated Swift
Aeronautes saxatalis | USA
FAMILY Trochilidae (hummingbirds)
369 species (c. 40–22 mya) [see also page 240]
SUBFAMILY Florisuginae (topazes & jacobins)
4 species [22–18.5 mya]
White-necked Jacobin
Florisuga mellivora | Ecuador
SUBFAMILY Phaethornithinae (hermits)
40 species [22–19.5 mya]
Grey-chinned Hermit
Phaethornis griseogularis | Ecuador
SUBFAMILY Polytminae (mangos & relatives)
29 species [18 mya]
Horned Sungem
Heliactin bilophus | Brazil
SUBFAMILY Lesbiinae (brilliants & relatives)
129 species [15.5 mya]
Golden-bellied Starfrontlet
Coeligena bonapartei | Colombia
SUBFAMILY Patagoninae (Giant Hummingbird)
1 species [14.5 mya]
Giant Hummingbird
Patagona gigas | Ecuador
SUBFAMILY Trochilinae (trochilines) 166 species [14 mya]
CLADE emeralds
112 species [12.5 mya]
Garden Emerald
Chlorostilbon assimilis | Costa Rica
CLADE mountain-gems
17 species [12 mya]
White-bellied Mountain-gem
Lampornis hemileucus | Costa Rica
CLADE bees
37 species [12 mya]
White-bellied Woodstar
Chaetocercus mulsant | Ecuador

Adaptations for Exceptional Lifestyles

All hummingbirds are bundles of highly strung, nervous energy and fly in an unmistakable way. In order to cope with their frenetic lifestyle, hummingbirds have evolved many unique anatomical and physiological adaptations, including possession of a higher metabolic rate than that of any other living vertebrate—they are truly in a league of their own in the animal kingdom. It is these attributes that enable them to thrive in so many different habitats, from sea level to the high Andes, from Alaska to the southern tip of South America.

Size, shape and structure

Most species of hummingbird are in the range of 6–12 cm (2.4–4.7 in) in length, often including a long bill and/or a long tail. The majority tip the scales at less than 6.5 g (0.23 oz), although a few larger species weigh around 12–14 g (0.42–0.49 oz), and the aptly named Giant Hummingbird weighs an exceptional 19–21 g (0.67–0.74 oz), which is about two-thirds the weight of an Eastern Bluebird or a House Sparrow (but quite a lot heavier than a typical small warbler). To be more specific, it would take two average small hummingbirds to balance the scales with a Yellow Warbler and, at the bottom end of the range, it would need *five* Bee Hummingbirds, the smallest of all birds at a mere 1.6–2.0 g (0.06–0.07 oz), to equal a small warbler, each one weighing only as much as just half a teaspoon of sugar.

Hummingbirds have a slightly elongated body that tapers back from broad shoulders. The head is relatively large and rounded, on a short but flexible neck. The feet, although tiny, are well suited for perching owing to the slim toes, three of which point forwards and one backwards, and the long, arched, sharp claws. As hummingbirds have exceptionally short legs, however, they cannot walk or move through vegetation in the way in which other birds do. Essentially, the only way they can get about is to fly from perch to perch, sitting still in a characteristically upright pose when at rest.

An obvious feature of a perched hummingbird is its long wings, folded back along the length of its body. Most birds have rows of wing-coverts that lie on top of, and hide, most of the flight feathers when the wings are closed, the outer half of the wing tucking up beneath them, out of sight. In contrast, the outer part of a hummingbird's wing is exposed as a long, curved 'blade', with the small wing-coverts covering only a small area towards the front. The bulk of the visible wing is made up of the primary feathers: ten long, stiff flight feathers that comprise up to 75% or more of the wing surface. For hummingbirds, it is these long outer flight feathers, which are dramatically overdeveloped relative to the rest of the wing, that do most of the work. The entire flight apparatus is powered by a group of strong muscles that sit close to the body. These highly specialized adaptations are a key feature of hummingbirds' anatomy, reflecting the importance of flight to their survival.

Remarkable senses and a unique anatomy

Hummingbirds' eyes are small and the acuteness of their vision unexceptional but they do, like most other birds, have a remarkable ability to perceive color. While mammals, including humans, have three kinds of color-receptor cones in each eye, attuned to red, green and blue light, birds have a fourth, allowing them to perceive some ultra-violet, beyond a human's color range. Humans can see 'non-spectral' colors, such as magenta and purple (as opposed to violet), when both our blue and red (but not green) cones are stimulated. As well as purple, birds can can see other non-spectral colors—ultraviolet plus green, ultraviolet plus red, ultraviolet plus yellow, and ultraviolet plus purple. Although such colors are impossible for us to imagine, hummingbirds have proved to be ideal subjects to study as they respond so strongly to colors that reveal flowers with the richest

The weight of a single **Yellow Warbler** *Setophaga petechia* is equivalent to that of five **Bee Hummingbirds** *Mellisuga helenae*.

FACING PAGE **Black-tailed Trainbearer** *Lesbia victoriae* | Ecuador sources of nectar. This has enabled researchers to show that hummingbirds are clearly able to discriminate between two kinds of 'ultraviolet plus red' hues and can separate 'ultraviolet plus green' from pure green and pure ultraviolet, for example. What we humans see as a flash of 'magenta' on a hummingbird's throat may be perceived as a very different color by other birds. This is explained in more detail in the chapter *Color and Iridescence* (see *page 49*).

It had for long been widely believed that hummingbirds have no sense of smell, and only a slight sense of taste. As mentioned earlier (*page 11*), however, it is now known that hummingbirds' taste receptors are adapted to enable them to discern sweetness, and more recent studies have shown that they are able to discriminate between low-sugar and high-sugar solutions when feeding. Although there is some evidence that hummingbirds also have a sense of smell that is used to locate nectar-rich flowers, albeit only secondarily to sight, this aspect of their biology is currently rather poorly understood.

Ultimately, all these attributes are of no value unless a hummingbird's minute brain can, quite literally, make sense of it all. It has been found that a hummingbird's brain processes information in a different way from that of all other animals: hummingbirds live at high speed and their brain has to work at a super-high speed. They are supremely attuned to their uniquely high pace of life, and it is difficult, if not impossible, for us to understand how hummingbirds perceive the world around them. For example, while other birds and mammals are especially sensitive to movement on the periphery of their vision and can pick up whatever is coming up behind them, hummingbirds are sensitive to movement all around. This is a particularly important adaptation if you have to focus on feeding, hovering, avoiding predators, attracting a mate, seeing off potential competitors and, in the process, be able to fly forwards, backwards and upside down, all at a pace unmatched by any other bird.

In all types of bird, the skeleton has evolved to provide strength and rigidity with the minimal weight

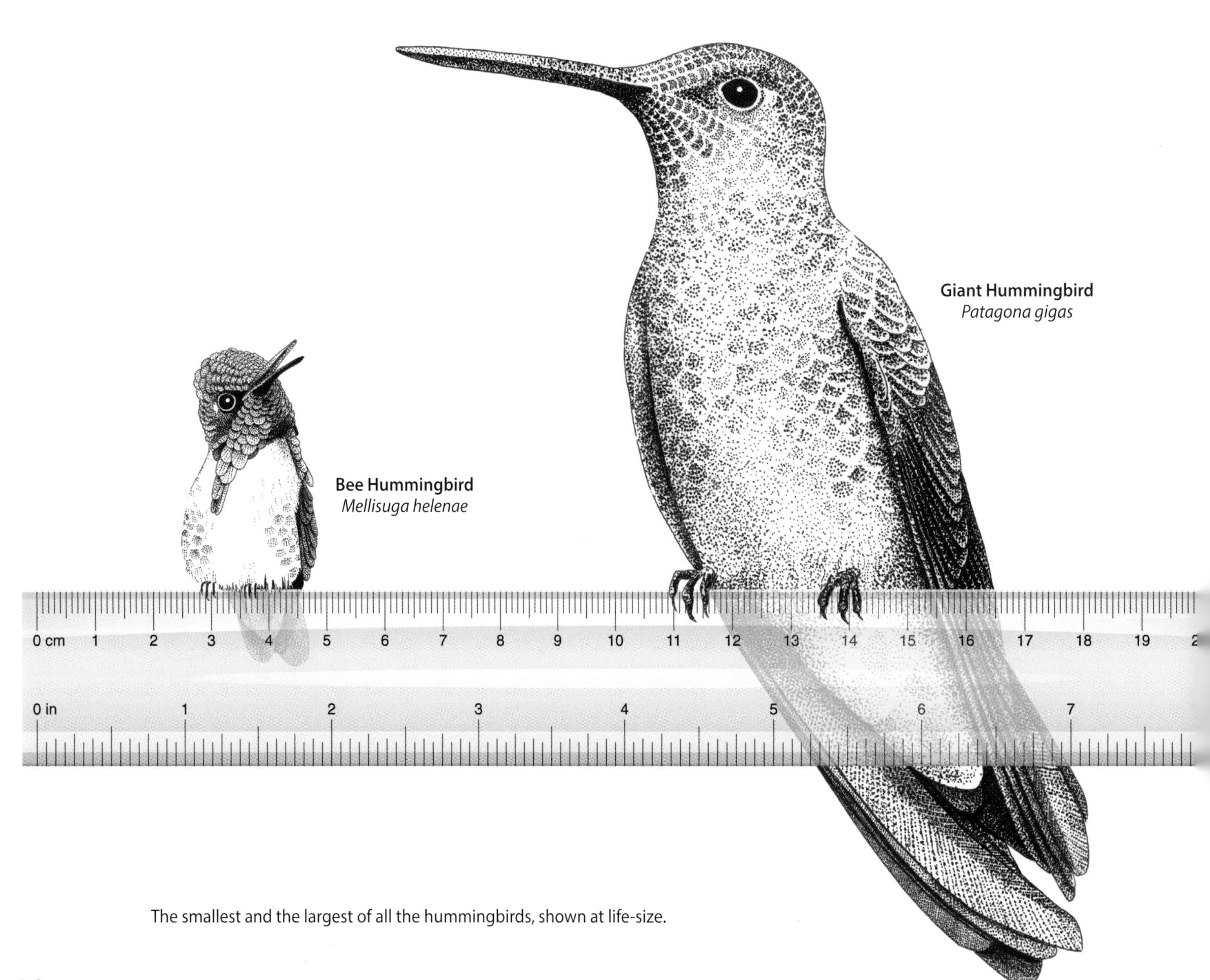

The smallest and the largest of all the hummingbirds, shown at life-size.

possible. This is the case especially with the relatively large skull, which, if it were overly heavy, would affect the bird's ability to fly efficiently. Hummingbirds, though, have had to face this challenge with the extra problem of their unusually small size. While many birds have hollow bones for lightness with internal 'struts' to provide the required strength, there simply is no room in a hummingbird skeleton for such structures. Relative to the rest of its skeleton, however, a hummingbird has larger wing and leg bones than most birds, and these, although seemingly so very thin and fragile, are indeed hollow, and its other bones are more or less porous. Both of these adaptations help to reduce weight considerably.

Prepared skeletons of hummingbirds in museums have an almost rice-paper-like, translucent quality, although most other details of the skeleton are relatively unremarkable. While the majority of mammals have seven neck vertebrae, birds can have up to 25, so hummingbirds, with just 14 or 15, are a little on the low side. A hummingbird skeleton does, however, have one characteristic feature above all others: a very large and deep, semi-porous sternum (breastbone). This is essential for the most extraordinary aspect of a hummingbird's lifestyle—the way it flies—since it provides the anchor to which the all-important flight muscles are attached.

Flight: taking it to extremes

Apart from their small size and dazzling colors, it is their manner of flight, and particularly their ability to hover, that epitomizes hummingbirds for most people. So, how do they do it? In short, it is all down to the shape of their feathers and the structure of their wings.

As with all flying birds, the feathers of a hummingbird have a central shaft with a flat vane each side. Each body (or 'contour') feather is near-symmetrical, but the flight feathers on a bird's wing are different, having a narrow 'outer' vane and broader 'inner' vane (or web), and a particularly stiff shaft in order to provide the necessary strength and rigidity. When the wing is closed, the feathers slide over one another; when it is spread, they open out like a hand of playing cards unfurled into a fan. The narrow outer vanes show on the upper side of the wing, with the outer or forward-facing edge exposed, while the broader inner vanes overlap beneath, with the inner edges revealed. As the wing is lowered the feathers create a single surface pushing against the air beneath, each inner vane pressed against the adjacent one. It is as the wing is raised, however, that the asymmetry of the feathers comes into play. Each stiff, narrow outer vane cuts upwards through the air, while the weaker, broader inner vanes are now twisted down by the pressure, allowing air to slip between the feathers. In this way, the downstroke pushes hard against the air, lifting the bird, while the upstroke lets air slip through, thereby minimizing the effect of being forced back down again.

Although hummingbirds are mostly very small, the density of their feathering is remarkably high. For example, a male **Allen's Hummingbird** *Selasphorus sasin* (*above*) has been found to have 1,459 feathers, and a female 1,659—a density five times greater than that found on most larger birds. | USA

A hummingbird's skeleton (*top*) is key to its survival, the very large sternum (breastbone) providing a strong anchor for the all-important flight muscles (*middle* and *bottom*). The tongue (*shown in red*), which is attached to the top of the head, wraps around the back of the head and can be greatly extended.

Quite a few birds, once airborne, are able to hover, either when searching for food or when displaying (terns, kingfishers, certain birds of prey and larks, for example), but it is hummingbirds that have truly perfected the art. Most birds that hover flap their wings more or less up and down. A hummingbird's wings, however, move in a very different and unique way, cutting forwards, leading-edge-first, and then rotating so that they are more or less upside down on the back stroke, the wingtips tracing a 'figure-of-eight' as they slice to-and-fro through the air. The degree to which a hummingbird's wings rotate varies according to circumstance: the bird's body may be almost horizontal with the head pushed forward, angled with the neck held upright, or almost vertical with the head and neck pushed forward. Altering the angle of the wing allows great agility and accurate positioning—and with it the ability to fly sideways, backwards and even upside down.

In order to fly in the way they do, hummingbirds have several anatomical features that are quite different from those of most other bird species. The majority of birds have rather long inner wing bones (or 'arms') and some movement in the carpal joint (or 'wrist'), the bend between the inner and outer parts of the wing. On a soaring bird such as an eagle or vulture, this enables the outer wing to be angled back, the feathers forming a single point, or pushed out straight, leaving the wingtip rounded or 'fingered' as the feathers spread apart. On a hummingbird, though, the innermost wing bones are very short and the carpal joint is stiff, so the wings are articulated only at the shoulder: in effect, the whole wing becoming one stiff, triangular blade. Most birds also have a small 'thumb' bone just beyond the carpal joint that is covered with a clump of feathers (the alula) which can be lifted to control air flow (this is obvious especially in pictures of birds of prey about to land). On hummingbirds, however, this thumb bone is reduced to a minuscule spike and, as a consequence, the front edge of their wings is perfectly smooth and 'cuts' efficiently through the air with the least possible resistance

The number of ribs that a hummingbird has is unusual, too: while most birds have six pairs of ribs, hummingbirds have eight. These are connected to the elongated, deeply keeled sternum that provides the essential firm attachment for the strong flight muscles. The front end of the sternum is joined to the humerus (equivalent to the human upper arm) by the coracoid bone. In most birds that have short wings, the humerus is extended outwards in flight, aligned along the wing, and simply moves up and down to power the wingbeats. In hummingbirds, however, the humerus points backwards, at right angles to the extended wing, and swivels, rotating halfway and back again, with the wing sticking out sideways. This increased rotation is possible only because of an unusually mobile shoulder joint, the coracoid bone having a ball-and-socket arrangement against the sternum, a feature found only in hummingbirds, swifts and treeswifts. In terms of energy efficiency, this adaptation is certainly advantageous, since the effort required by a hummingbird to sweep its wings back and forth while simultaneously beating with a swivel action is considerably less than that of a typical bird pushing its humerus up and down.

The action of a hummingbird's wings is fairly easy to understand, but the speed at which this takes place is mind-boggling. For example, small hummingbirds move their wings an average of 70 or 80 beats *per second* in normal flight. While this figure is in itself extraordinary, what is utterly astonishing are the figures for display flights, where occasional bursts of up to 200 beats per second have been recorded. When traveling from one place to another hummingbirds also fly at high speed, often around 50–80 kph (30–50 mph), with occasional bursts up to an amazing 95 kph (60 mph), fast enough to keep up with a car on the highway.

[continued on *page 21*]

The spectacular male **Long-tailed Sylph** *Aglaiocercus kingii* has a remarkably long tail, almost twice the length of its body. | Ecuador

Many hummingbirds are extremely agile in flight, as depicted in the illustrations *below*: as well as being able to hover and rotate on the spot, they have a unique ability to fly backwards or upside down, and can even perform a somersault to get away from a rival or predator.

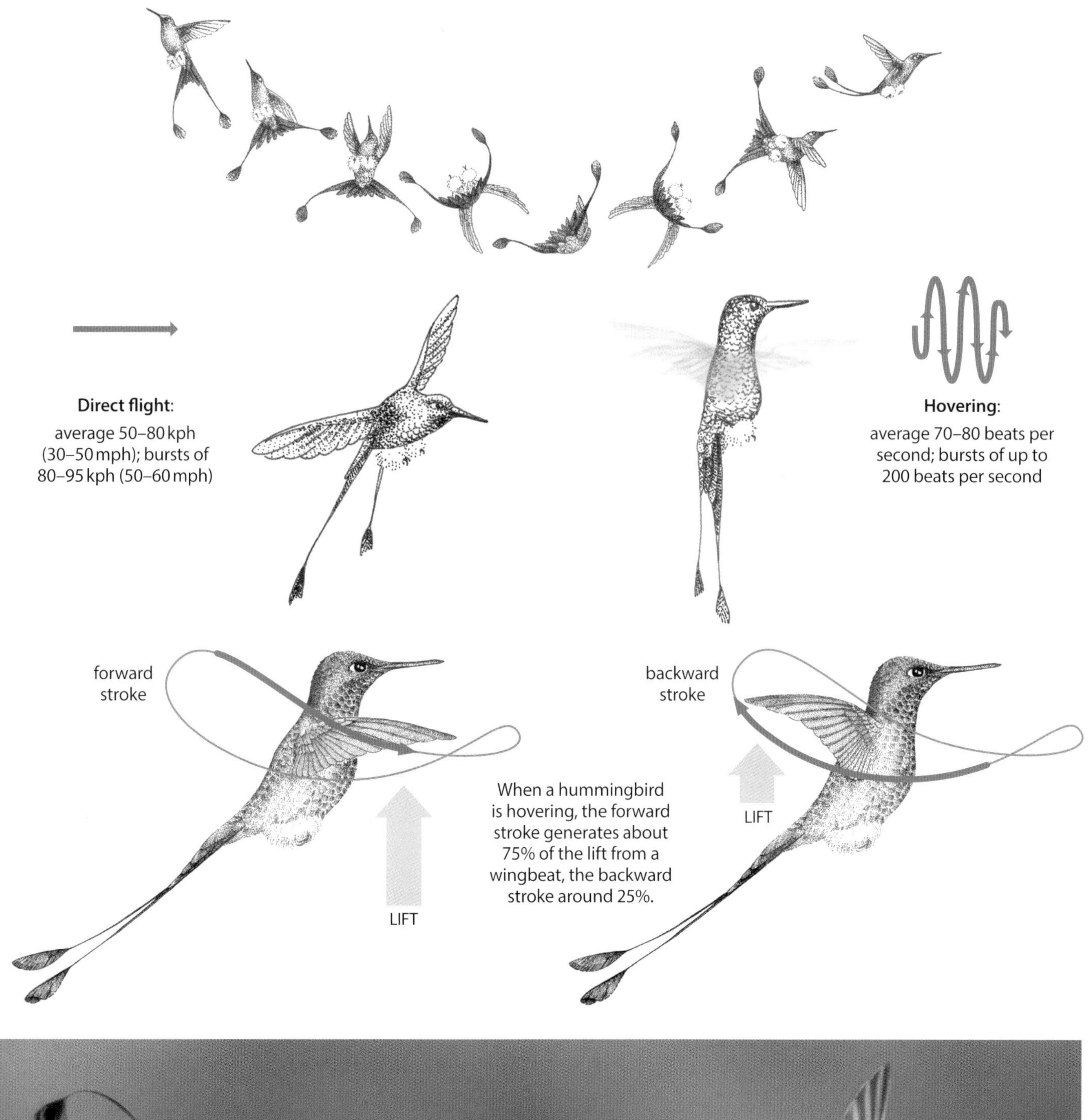

These photographs show different wing positions of a **Booted Racket-tail** *Ocreatus underwoodii* in direct flight. | Ecuador

Tails

The majority of hummingbirds have a short, rounded or square-ended tail, but the males of some species have long outer tail feathers that create a fork when the tail is spread, an extreme example being the Swallow-tailed Hummingbird (*illustrated below*). Although the shape of the tail is important in helping to maintain stability or enable tight, rapid changes of direction in flight, in a few species the males have tail feathers that are especially long and appear to be designed simply to impress the female—such as remarkable streamers, rackets and spatulas.

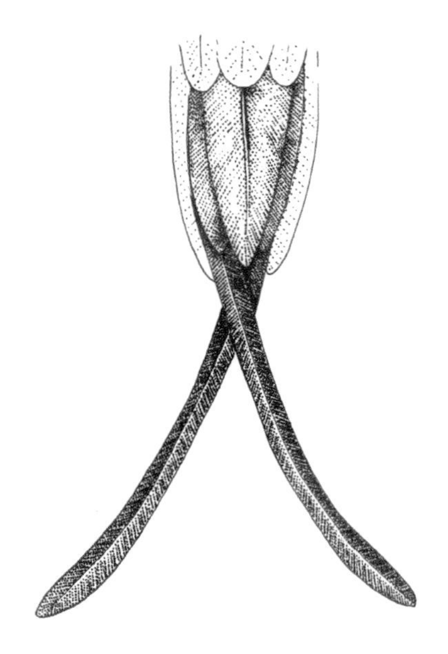

Crimson Topaz
Topaza pella

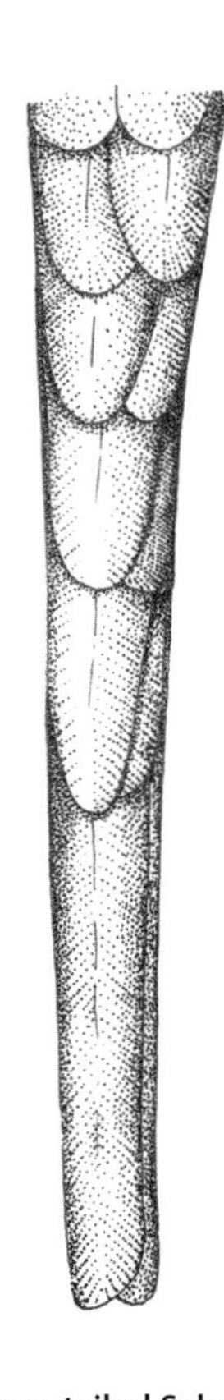

Long-tailed Sylph
Aglaiocercus kingii

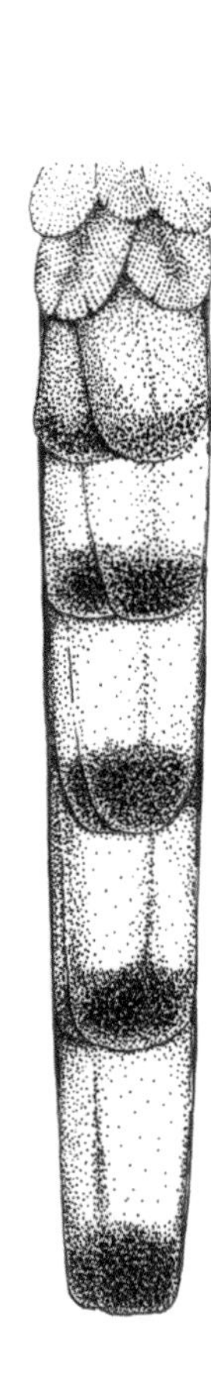

Red-tailed Comet
Sappho sparganurus

Red-billed Streamertail
Trochilus polytmus

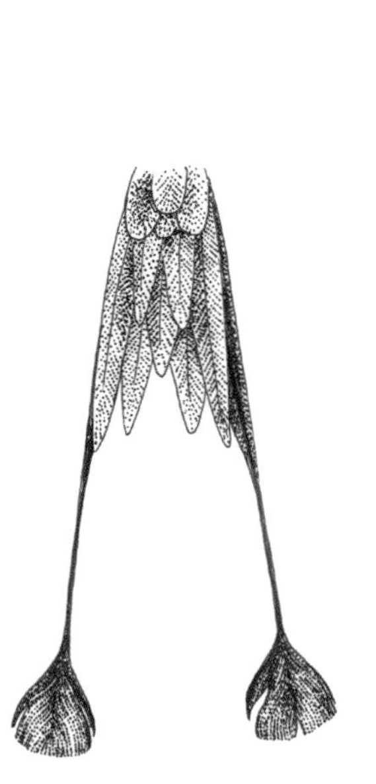

Booted Racket-tail
Ocreatus underwoodii

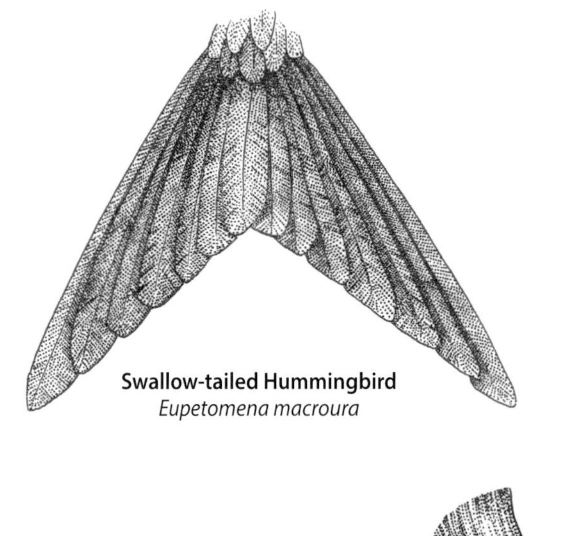

Swallow-tailed Hummingbird
Eupetomena macroura

Marvelous Spatuletail
Loddigesia mirabilis

One special feature of hummingbirds that allows this super-natural speed is the way in which their muscles have developed. The wing is pulled down by the large 'pectoralis major' muscle and up by an enlarged 'supracoracoideus', a strong muscle that binds the keel of the sternum around the coracoid to the base of the humerus. These two muscles account for up to 30% of a hummingbird's weight (for most bird species this figure is around 20%). While other birds have a strong downstroke and weak upstroke, hummingbirds' powered upstrokes have an almost equal importance, allowing the aerobatic and all-round directional ability that seems so effortless. A hummingbird's control while its wings are whirring away at an unimaginable speed is also quite extraordinary. When it is hovering, either it can keep its head fixed absolutely still in relation to the ground or foliage, or the body can remain still while the head moves slowly up and down to follow minor movements of flowers in the breeze.

An inevitable consequence of all this high-speed flapping is a demand for huge amounts of energy. Perhaps paradoxically, though, the relatively large surface area of a hummingbird's wings in comparison to its body means that the bird loses heat especially quickly and the energy demand is therefore even greater. When hovering, however, a hummingbird is able to compensate by using the heat produced by its rapidly expanding and contracting flight muscles to regulate its body temperature—and can do so far more precisely than any other bird can. In cold conditions, wingbeats are shallower and as a consequence heat production is reduced, although this is offset by increasing the frequency of wingbeats. In thin air at high altitude, the wingbeats are deeper in order to counteract the reduced air pressure against the surfaces. As explained later in this chapter (*page 28*), the action of so many muscles expanding and contracting so fast also helps to pump oxygen around the body when it is most needed.

Some hummingbirds have a long or extravagantly shaped tail, which, as well as being decorative, has a very important function in flight. A large surface area, and frequently a forked shape, not only gives greater directional control but also provides extra lift and balance, especially in the abrupt change from forward to hovering flight. In addition, the pointed sides of a long, graduated or forked tail are particularly effective aerodynamically in reducing drag. There is some variation among different types of bird in the number of tail feathers they possess, although most have 12. With a single exception, all hummingbirds have ten tail feathers, the Marvelous Spatuletail being the one that breaks the rule by having just four feathers in its specially modified and spectacular tail.

A **Green-breasted Mango** *Anthracothorax prevostii* hovering, with its rather rounded tail spread and its wings twisted as they retract from a forward stroke. | Costa Rica

Flight

Photographs taken with high-speed flash can capture hummingbirds in extraordinary poses and, as illustrated here and on the following pages, show their tremendous control when maneuvering to extract nectar from a flower and then backing away.

RIGHT The wings of a **White-bellied Woodstar** *Chaetocercus mulsant* beat so fast that they appear as a blur when the bird is hovering. | Ecuador

BELOW Although all hummingbirds have strong flight feathers, the shafts of the outermost primaries of male sabrewings, such as this **Napo Sabrewing** *Campylopterus villaviscensio*, are unusually thick and curved (hence the word 'sabre' in its name)—but the reason for this is unclear. One possibility is that they are an adaptation to aerial foraging for insects, aiding tight turns and twists when searching for food; another is that the feathers are strengthened to prevent damage when flying through forest. Other hummingbirds that live in the same environment, however, seem to manage perfectly well without this feature, as do female sabrewings. A third suggestion is that the feathers may have some importance in display or female choice of the fittest mate, perhaps producing a special sound (although this has not been noted), or aiding a particularly impressive, aerobatic flight. This is just one of many mysteries surrounding hummingbirds that has yet to be solved. | Ecuador

ABOVE **Brazilian Ruby** *Clytolaema rubricauda* | Brazil BELOW **Velvet-purple Coronet** *Boissonneaua jardini* | Ecuador

ABOVE **Broad-billed Hummingbird** *Cynanthus latirostris* | USA BELOW **Black Jacobin** *Florisuga fusca* | Brazil

ABOVE **Black-bellied Hummingbird** *Eupherusa nigriventris* | Brazil BELOW **Andean Emerald** *Amazilia franciae* | Ecuador

ABOVE **Great Sapphirewing** *Pterophanes cyanopterus* | Colombia BELOW **Crowned Woodnymph** *Thalurania colombica* | Ecuador

ABOVE **Marvelous Spatuletail** *Loddigesia mirabilis* | Peru BELOW **Buff-winged Starfrontlet** *Coeligena lutetiae* | Ecuador

Cardiovascular champions

The unique way in which hummingbirds fly, involving constant exertion, requires very specific physiological adaptations. As hummingbirds are warm-blooded, they require much more energy than large insects with a similar feeding style do. The heart of any flying bird works remarkably hard, yet in a small songbird it may constitute a mere 1.3% of the body weight (compared with about 0.5% in a human); a hummingbird's heart, however, can be up to 2.5% of its body weight. While a human's heart may beat around 50–70 times per minute, or double that when working hard or under emotional or physical stress, a hummingbird's heart achieves an amazing 500–600 beats per minute, even when it is perched. Even more incredible is that a hummingbird's heart rate may exceed 1,000 beats per minute in high-speed flight, when it is powering blood around the body to keep the muscles working.

Flight, and hovering in particular, requires not only a fast heart rate but also very efficient use of oxygen. To get that all-important oxygen into the blood, a human takes about 15 breaths per minute and a pigeon 30 breaths per minute. A hummingbird, on the other hand, may take an astonishing 300 breaths per minute at rest and 500 per minute when hovering: that is eight gulping breaths every second! Birds, however, do not use lungs in the same way as mammals do: the lungs are relatively small and rigid and do not pump in and out like those of mammals. Instead, oxygen is pushed through the body by a series of air sacs, the process being activated by muscle movement. This ensures that the bird pumps air just when oxygen is most needed, when the muscles are working hard. Oxygen is extracted in the bronchial structures within the lungs—millions of microscopic tubes with honeycombed walls that are almost inconceivably minute.

Another adaptation that reveals the vital importance to hummingbirds of the efficient use of oxygen is that they have the greatest concentration of hemoglobin—oxygen-carrying red blood cells—of any animal, and indeed their hemoglobin has enhanced oxygen-binding properties. This explains how, even at high altitudes where the oxygen content of the air is reduced, hummingbirds are able to hover with only a slight increase in oxygen consumption, although there may be a reduction in the duration of hovering bouts.

Surviving the cold

Hummingbirds are found throughout the Americas wherever there are flowers that provide them with nectar on which to feed. In mountainous regions, hummingbirds may occur at altitudes of up to 5,000 m (around 16,400 ft) above sea level, where they have to contend with temperatures changing by as much as 15°C (27°F) during every 24 hours. In such situations, rather than feeding regularly, they take in energy frantically in short bursts and then reduce their energy consumption during long periods of inactivity. While a hummingbird's normal body temperature is around 40–42°C (104–108°F) (about the same as that of other

In North America, where 14–16 hours of daylight are available for feeding during the summer, torpor is a strategy only rarely used by hummingbirds. The short nights are usually easy to survive with the energy that a bird is able to store each day, although some species, such as **Anna's Hummingbird** *Calypte anna* (*left*), will become torpid if the temperature falls below freezing. | ♀, Canada. The **Black Metaltail** *Metallura phoebe* (*right*) holds the record for the lowest recorded body temperature of any non-hibernating organism. | Peru

birds but higher than our own), hummingbirds also have another, extra-special ability—being able to reduce their metabolic rate and go into a state known as torpor. This is akin to hibernation, albeit for only a relatively short period of time, the birds adjusting their body temperature and metabolism to local conditions. Remarkably, they are able to reduce their body temperature and maintain it at around 5–10°C (41–50°F). This allows their heart rate to fall to a mere 50 beats per minute, and their breathing rate may be reduced by half. Overall, they can reduce their metabolic rate by anything up to 95% for a few hours. As a consequence, the bird becomes lethargic and unresponsive, but going into torpor and thereby conserving large amounts of energy is usually enough to get it through most long, cold nights. In fact, the record for the lowest body temperature ever known in birds or non-hibernating mammals is held by a hummingbird—a Black Metaltail—at 3.26°C (37.87°F). To put this into context, in humans a drop in body temperature of a mere 2°C induces hypothermia!

Torpor does, however, have its risks: predators may find a sleepy hummingbird easy prey as the bird cannot simply leap into life at a normal pace. Moreover, in some instances, especially during particularly cold periods when temperatures fall below 15°C (59°F), the bird's energy reserves may fall too low for it to be able to recover. But despite these risks, hummingbirds will sometimes also go into a state of torpor overnight when on migration in order to conserve energy for the long flight the next day.

Sitting pretty: feather care

Feathers are of paramount importance to all birds and to keep them in good condition a bird will preen, using its bill to clean its feathers, spreading oil from a preen gland at the base of the tail to keep them 'conditioned', and to 'zip' any gaps back together again. Many hummingbirds, however, have a very long, thin bill and despite being able to stretch and contort their neck have great difficulty reaching all their feathers with their bill tip. Instead, they often 'preen' using their claws; despite having such tiny legs a hummingbird is able to lift its feet up inside the folded wing in order to reach its head, neck and some body feathers. The Sword-billed Hummingbird, has such a long bill that it cannot be used at all for preening, and the bird has no alternative but to do the job as thoroughly as possible with its extra-long claws.

Hummingbirds will also bathe to help rid the feathers of dust and parasites. Some will hover over water, dipping in breast-first, or rising and falling to immerse their tail feathers, occasionally settling briefly on the surface to force water through their plumage. Others will dive and fully submerge in forest streams. More often, however, they tend to bathe during rain showers, in water dripping from foliage, use the fine spray from a tiny splashing waterfall, or rub against wet foliage or moss. In whichever way they achieve it, most hummingbirds bathe several times a day, perhaps as an alternative to the thorough preen that some may not be able to accomplish properly.

Hummingbirds, such as this **White-chinned Sapphire** *Amazilia cyanus* (*left*), preen regularly to keep their feathers in good order. | Brazil
Others, such as this **Festive Coquette** *Lophornis chalybeus* (*right*), are able to lift their feet to reach their neck in order to 'preen'. | Brazil

LEFT A **Velvet-purple Coronet** *Boissonneaua jardini* bathing by forcing its body into wet moss. | Ecuador; RIGHT A **Glittering-throated Emerald** *Amazilia fimbriata* taking a bath by fluttering on top of a wet leaf. | Brazil; ; BELOW A **Brazilian Ruby** *Clytolaema rubricauda* bathing by spreading its tail and one wing, and then the other wing, during a light rain shower. | Brazil

Bathing

In order to ensure effective feather-care, hummingbirds bathe frequently. They achieve this in a variety of ways, forcing water through their plumage by splash-diving into streams, perching under tiny waterfalls, spreading their wings and tail during a rain shower, or rubbing against a wet leaf or water-soaked moss.

ABOVE **Crowned Woodnymphs** *Thalurania colombica* often bathe by hovering over water and dipping on to the surface, sometimes submerging briefly. | Costa Rica

BELOW Some hummingbirds, such as this **Anna's Hummingbird** *Calypte anna*, occasionally sunbathe, fluffing out their feathers and spreading their wings. This behavior probably helps to keep the feathers in good condition and drive parasites from the bird's plumage; it may also provide a means of regulating body temperature. | USA

LEFT A **Steely-vented Hummingbird** *Amazilia saucerottei* in heavy molt. | Colombia; RIGHT a **Brown Inca** *Coeligena wilsoni* with pollen grains on its bill and head. | Ecuador

Molt—the process of replacing feathers

No matter how well a bird looks after its feathers, they will gradually wear out and need to be replaced, a process known as molt. Most small birds have both a complete molt (in which all the feathers are replaced over several weeks) and a separate partial molt (whereby only the smaller body feathers are replaced) each year. Once again, hummingbirds have a unique strategy involving a single, complete annual molt that starts soon after breeding and continues over several months. This prolonged molt reduces the risk of gappy, inefficient wings caused by too many feathers being missing at any one time. Most birds have ten main flight feathers (primary feathers, or primaries) on each wing that are replaced in sequence from the innermost outwards, but in the case of hummingbirds the outer two are shed and regrown in reverse. It has been suggested that this may be in order to avoid a temporary gap inside the leading edge of the wingtip which would disrupt hovering flight—although such a gap would appear regardless of whether the second outermost primary was molted ninth or tenth in the sequence. Another possibility is that the long, heavy ninth primary is given better support if it grows next to a new and strong outermost feather rather than an old, worn one. Asymmetry between left and right wings, and indeed tail sides, is frequent in molting hummingbirds, although they are clearly well able to compensate in flight.

The body feathers are molted within the duration of the primary-feather molt, but in many species of hummingbird the iridescent head and throat feathers are molted late in the sequence and are therefore fresh and bright immediately before the next breeding season. Within any one species, the timing of molt may vary from place to place, suggesting that it is adjusted to local conditions and ensuring optimal suitability for breeding, rather than being fixed within a specific time period. One way or another, hummingbirds look after their plumage well and, when it matters, are able to show off their stunning colors and patterns to best effect.

Living on nectar

Hummingbirds are renowned for having perfected the art of feeding from flowers—but why flowers, and why is it that wherever in the Americas there are flowers you are likely to find hummingbirds? It all comes down to a hummingbird's need for a very high-energy food to power its all-essential flight muscles. Although hummingbirds do occasionally eat a small amount of pollen and a few invertebrates, their diet comprises almost entirely (about 90%) nectar, which is particularly rich in sugars and carbohydrates. In fact, a hummingbird, in order to generate sufficient energy, may have to feed from hundreds of flowers, even as many as 2,000, every day: a remarkable statistic that says much about the abundance of flowers (even if only tiny florets), as well as the ability of hummingbirds to exploit them. But this association is also equally vital to the flower. As with many insects, and some mammals such as bats, when a hummingbird feeds on nectar it inadvertently transfers

FACING PAGE A **Long-billed Hermit** *Phaethornis longirostris* feeding from a *Heliconia* flower: look closely and you will see pollen grains falling on to the bird's head. | Costa Rica

pollen from the anthers of one flower to the stigma of the next, resulting in fertilization; indeed, certain plants rely entirely on hummingbirds for pollination.

The close link between hummingbirds and flowers is no better illustrated than by the coevolution of flower shape and bill shape to ensure a perfect 'fit'. The length of a flower's corolla (tube of partly fused petals) can often determine which species of hummingbird is able to feed from it. The majority of hummingbirds have a long, thin bill, either straight or downcurved, that is 'made to measure' for particular types of flower. The longest-billed species tend to visit more widely dispersed flowers with longer corollas, repeating a regular circuit around known blooms, giving each one time to create more nectar before the next visit, a feeding strategy known as traplining.

Hummingbirds that have a relatively short bill generally occupy smaller feeding territories, taking nectar from flowers that are clustered together, although they are frequently disrupted by larger, more powerful hummingbirds that are able to withstand the aggressive responses of the territory-holder. Some of the especially small, more insect-like hummingbirds such as the woodstars and coquettes, however, are able to visit the feeding territories of other species unnoticed, or feed on minute flowers that larger species are inclined to ignore. Some short-billed hummingbirds (such as the thornbills and wedge-billed hummingbirds) will pierce a flower's corolla to 'steal' nectar, while others reach nectar by probing through holes made at the base of flowers by other 'nectar thieves' such as flowerpiercers (a group of birds with a distinctively hook-tipped bill that is adapted for this purpose). Even flowers with a short corolla that produce relatively little nectar may still be visited by hummingbirds, but mostly only at certain times of the day—and usually only by short-billed species with a more generalized feeding strategy.

Although most flowers are accessible to several species of hummingbird that have a similar bill structure, in some cases hummingbird and flower have clearly evolved in parallel. The two species of sicklebill are an extreme example, each having a remarkably long, sharply downcurved bill that is a perfect match for the corolla-tubes of certain species of *Centropogon* (a genus of plants that occurs from central Mexico to tropical South America) and some *Heliconia* species. They have relatively short wings, and rather than hovering to feed, as do most hummingbirds, they have a tendency to perch, probably because of the difficulties of maneuvering their awkwardly shaped bill deep into the flower.

Some hummingbirds steal nectar by making a hole at the base of a flower, whereas others take advantage of holes made by birds such as this **Slaty Flowerpiercer** *Diglossa plumbea* | Costa Rica

As mentioned previously, hummingbirds, have remarkable color vision and seem well able to judge a flower's suitability for feeding by its color. Many flowers have changing patterns of color that are revealed only in ultraviolet light. While this is invisible to us, the prominent, but not especially large, eyes of a hummingbird are able to detect ultraviolet signals that indicate the periods when large amounts of nectar are available; this is generally at the time when the flower is most fertile. Nevertheless, despite being so accurate in their selection of nectar-rich blossoms, hummingbirds seem continually to be exploring, trying to discover new feeding opportunities, and will frequently investigate other natural or even unnatural objects, especially if they are red or orange. A red flower may not be selected for any particular characteristic other than the fact that it is simply easier to see against the contrasting greens of a forest backdrop. Even to our eyes, bright red can tend to 'leap out' against a strong green background. Anyone who has watched a hummingbird feeding in the wild can attest to the fact that its vision seems to be especially sensitive in the red wavelengths; indeed, hermits will often make a disconcertingly close inspection of a small item of red clothing, such as a bandana or cap.

Hummingbirds have excellent memories for individual flowers and patches of blooms that consistently yield abundant nectar, typically favoring blooms that are more or less tubular in shape and have little or no scent. In some instances the corolla-tube may be particularly thick, probably as a defense against those birds that try to steal nectar without actually pollinating the flower. Nectar production in the flowers of some individual plants may be staggered, offering a constant supply of food for

FACING PAGE A denizen of Andean montane forests, the **Sword-billed Hummingbird** *Ensifera ensifera* is aptly named. Not only does it have the longest bill of any hummingbird, at 9–10 cm (3.5–4.0 in), it is also the only bird in the world to have a bill longer than its body. Having such an exceptional bill enables it to access the nectar of flowers with an extremely long, tube-like corolla, such as this angel's trumpet *Brugmansia* (which can have corollas up to 12 cm/5 in in length)—a remarkable foodplant–hummingbird relationship and an amazing example of coevolution. | Ecuador

A bird of humid deciduous forests from Panama to Peru, the **Violet-bellied Hummingbird** *Amazilia julie* must hover to extract nectar from a variety of drooping, tubular blossoms, using its long, straight bill as a probe. | Ecuador

The **Collared Inca** *Coeligena torquata* is a common and often rather conspicuous hummingbird of humid Andean forests. It feeds from many plants, particularly those with narrow, tubular flowers that allow only long-billed hummingbirds access to nectar. | ♀, Ecuador hummingbirds. The benefits of the association between hummingbirds and plants, however, is sometimes weighted more towards the plant than towards the bird. For example, certain plants produce a mixture of flowers, of which some have nectar and some do not. This not only saves the plant energy but also forces the hummingbird to visit a wider selection of blooms in order to obtain the food needed to sustain itself, thereby increasing the chances of pollination.

Plants produce nectar for one reason alone, and that is to attract pollinators and ensure fertilization. Although this requires the expenditure of energy, the costs to the plant is undoubtedly outweighed by the benefits. The result is a fascinating 'cat and mouse' relationship, the shapes both of flowers and of hummingbirds' bills continually coevolving—a process that continues to determine the abundance and distribution of both.

Extracting nectar

Hummingbirds extract nectar from a flower by using their long, sensitive tongue that is sheathed in a thin bill. The bill can be opened wide but is flexible enough to be opened only towards the tip when feeding. The tongue curls around the back of the skull when retracted, much like a woodpecker's (see *page 17*).

It had for long been thought that a hummingbird's tongue 'sucked' up nectar with rapid movements at the tip, rather like a person using a drinking straw. More recent studies, however, have shown that the technique is a little more complex. To extract nectar quickly, the tongue of most species is typically long and split lengthwise, creating two curled 'troughs' or grooves with brush-like points. The tip of the tongue can be pushed far forwards and is flicked in and out of a flower up to 13 times per second. This flicking, or 'licking', action forces nectar along the tongue through the tiny grooves. The bill itself may be closed momentarily to squeeze nectar from the tongue into the throat between each flick, with a rapid gobbling action.

As well as finding and feeding at many flowers each day, hummingbirds also need to process all of this energy-rich food efficiently within their body. In order to do so, they have yet more incredible adaptations. A crop full of nectar can be emptied in four minutes, and a hummingbird's small intestine can extract 99% of the glucose from the nectar within 15 minutes. The necessary digestion in the intestine, however, takes time and, although the fastest of any animal, this limits the intake of food. As a consequence, hummingbirds must rest between feeding bouts to digest, thereby conserving energy. But they cannot rest for too long as they must feed regularly if they are to obtain sufficient nectar to survive. To achieve this, many hummingbirds take their first sip of nectar before dawn and continue to be active until late in the gathering dusk. To put this into context, a human processing as much energy as a hummingbird would need to eat around 170 kg (370 pounds) of potatoes each day! Overall, most hummingbirds spend around 30% of their day in feeding and the remainder in resting, singing, preening and bathing.

The tongue of all hummingbirds, as illustrated by this **Anna's Hummingbird** *Calypte anna*, can be extended well beyond the tip of the bill: an essential adaptation enabling it to reach and extract nectar from flowers. | Canada

This array of hummingbird heads illustrates the tremendous variety of bill sizes and shapes, each adapted for particular types of flower. Short-billed hummingbirds can probe deeply into flowers with a short corolla, while the nectar in long, narrow flowers can be reached only by those species with a long bill. Some hummingbirds have a hook-tipped or toothed bill—both believed to be adaptions for feeding on invertebrates and perhaps also for defending a territory and/or preening.

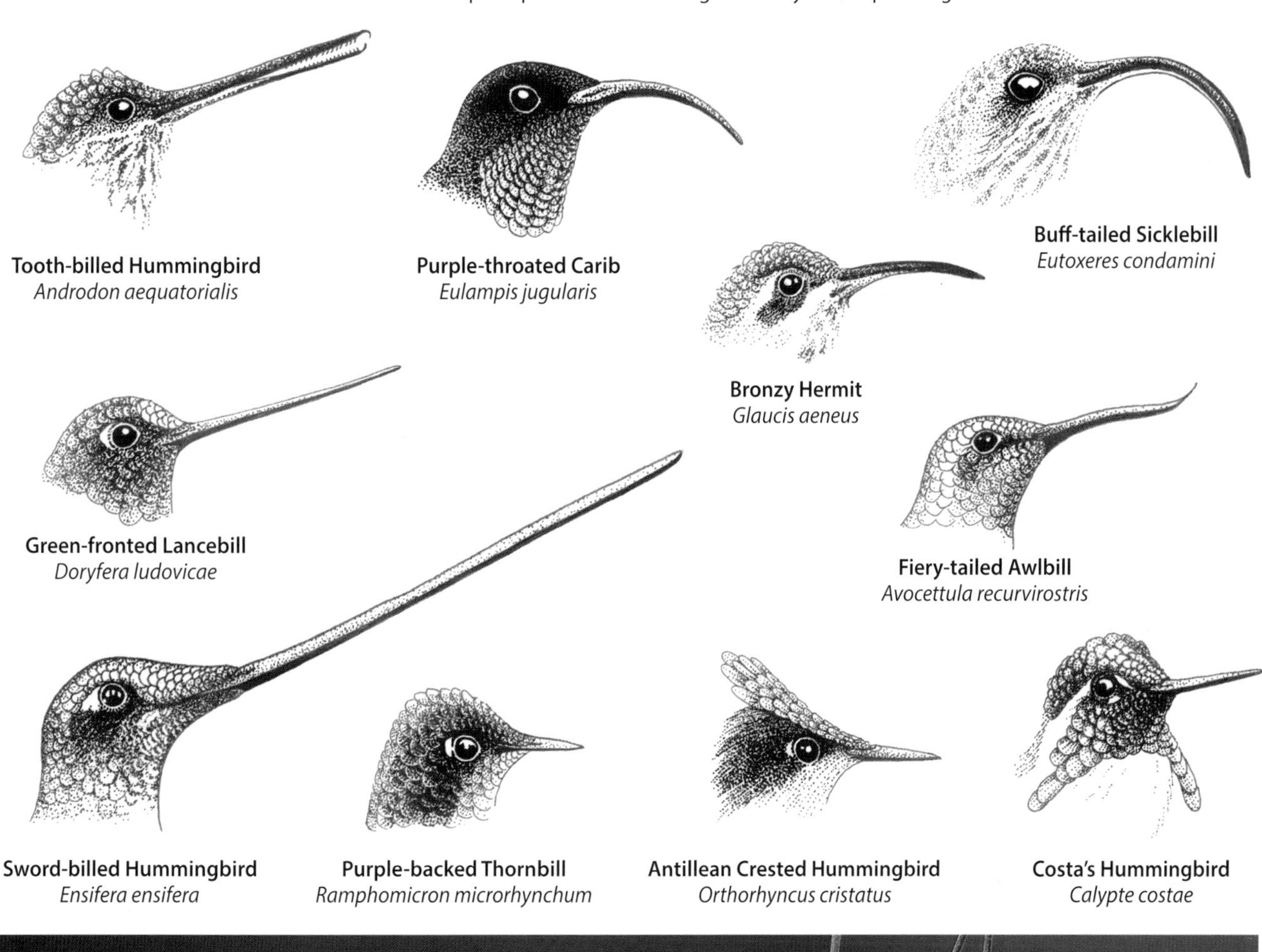

A number of hummingbirds are habitual nectar thieves, probing the base of flowers and therefore not fulfilling the role of pollinator. The bill of the two wedge-billed hummingbirds is specially adapted for this purpose, the tip being sharply pointed—and hence their alternative name of daggerbills. This is a **Western Wedge-billed Hummingbird** *Schistes albogularis*. | Colombia

Bills and feeding

As shown on the previous page, hummingbirds exhibit a tremendous variety of bill shapes and structures, reflecting their close association with a range of different flowers. In some cases, hummingbird and flower appear to have coevolved, leading to specializations that now force a hummingbird with a very long or very strongly curved bill to feed from, and therefore potentially pollinate, just one or a few particular types of flower. Many hummingbirds, however, have a a relatively short 'all-purpose' bill which enables them to feed from a much wider range of blossoms, or even to pierce the base of the flower to access the nectar.

FACING PAGE The **White-tipped Sicklebill** *Eutoxeres aquila* has a sharply downcurved bill that is perfectly adapted for feeding from a limited range of plants that have long, curved flowers, such as this *Heliconia*. | Peru

ABOVE Most short-billed hummingbirds, such as the **Purple-backed Thornbill** *Ramphomicron microrhynchum* (*left* | Colombia) and **White-crested Coquette** *Lophornis adorabilis* (*right* | Costa Rica), have a rather generalized feeding strategy, taking nectar from a wide range of small flowers.

BELOW Some hummingbirds, such as this **Reddish Hermit** *Phaethornis ruber*, have the right-sized but wrong-shaped bill to feed from certain flowers and sometimes have to be creative in order to reach the nectar! | Brazil

Not all hummingbirds hover to feed: many perch or cling onto vegetation close to the flower, thereby conserving energy. Examples of species that often feed in this manner include **Green-crowned Brilliant** *Heliodoxa jacula* (♀, *above left* | Costa Rica), **Velvet-purple Coronet** *Boissonneaua jardini* (*above right* | Ecuador), **Tourmaline Sunangel** *Heliangelus exortis* (*below* | Ecuador), and **Shining Sunbeam** *Aglaeactis cupripennis* (*facing page* | Colombia).

Anna's Hummingbird *Calypte anna*

Chestnut-breasted Coronet *Boissonneaua matthewsii* | Ecuador

Dietary supplements

In addition to nectar, hummingbirds need to eat insects and other small invertebrates (arthropods) in order to obtain protein (which nectar lacks) and calcium for eggshell production. Females supplement this small amount of calcium intake by licking up bits of soil, sand, or even wood ash, as they hover over the forest floor. Insects are generally caught in the air, especially by short-billed and straight-billed hummingbirds; species that have a very long or curved bill are unable to do this and instead take insects and other small invertebrates by picking them from leaves. Unlike nectar, which is diverted straight to the intestines, where it can be processed rapidly, arthropod food is processed in the stomach, the hard parts being ground in the gizzard.

Like the swifts and treeswifts, their closest relatives (see *page 12*), hummingbirds are able to see insects at close range and catch them in their bill or wide-open mouth in flight. In order to achieve this, they have evolved specially adapted neck muscles, allowing a fast and accurate sideways movement of the head, as well as a quick snap of the bill, both of which are essential for this feeding technique to be effective. Short-billed hummingbirds can open their gape particularly wide (as shown by the Chestnut-breasted Coronet, *above*) and take in flying insects at the back of the mouth, rather than catching them in their fine and delicate bill. In North America, some hummingbirds have adapted to visiting the holes made in tree bark by sapsuckers (fairly small woodpeckers), feeding not only on the resulting sap runs but also on the insects that these attract. Ruby-throated Hummingbirds, for example, may do this before the full range of nectar-bearing flowers is in full bloom in spring (see also *Epic migrations, page 174*).

The various species of hermit, which live in the understory beneath a dense forest canopy, will sometimes hover around spiders' webs, snapping up not only the spiders' catches but also the spiders themselves. The Saw-billed Hermit (*facing page, top right*) has a particularly heavy, thick bill with a hooked tip that appears to be an adaptation to feeding on such invertebrates. Similarly, the uniquely serrated bill of the Tooth-billed Hummingbird (*facing page, top left*) is believed to be an adaptation to catching insects and other invertebrates, although it may also have a function in territorial defense and possibly when preening.

'Typical' hummingbirds 'flycatch' more frequently than do hermits, hawking insect prey in the air. Brown Violet-ear and occasionally Anna's Hummingbird (illustrated *above left*), for example, feed in this manner, catching insects by sallying out from a perch like a flycatcher, or by hovering for longer periods and darting to-and-fro, in effect using the hover as its 'perch'. Another species that is especially known for eating invertebrates is the Fiery-tailed Awlbill (*facing page, bottom*), which has a slender bill that, unusually for a hummingbird, tilts upwards at the tip. Although it feeds mostly on nectar, taken from many flowers along regular routes, it will also take insects, either hawking them in flight or picking them or other invertebrates from the undersides of leaves, this perhaps explaining its unique bill shape.

ABOVE LEFT **Tooth-billed Hummingbird** *Androdon aequatorialis* | Colombia; ABOVE RIGHT **Saw-billed Hermit** *Ramphodon naevius* | Brazil;
BELOW **Fiery-tailed Awlbill** *Avocettula recurvirostris* ♀ | Colombia

El Jardín Encantado in the village of San Francisco de Sales, at an altitude of 1,500m (4,900 ft) just 47km (29 miles) from Bogotá, the capital city of Colombia, is an amazing hummingbird spectacle.

Hummingbirds and feeders

Hummingbirds' almost total dependence on sugary food means that they are readily attracted to special feeders that allow them to sip sugar-rich solutions in place of nectar. In many instances, these feeders may attract just a handful of hummingbirds, but in some cases the numbers visiting during the course of a day may be quite amazing. One spectacular example is El Jardín Encantado (the 'Enchanted Garden') in the village of San Francisco de Sales, near Bogotá, in Colombia, where 40 feeders are maintained and 27 species of hummingbird have been recorded. The total number of individual birds visiting is estimated to be more than 1,000 each day, consuming up to 450 kg (1,000 lbs) of raw sugar every month! As a general rule, it has been suggested that the normal number of birds visible at any one time at an individual feeder or group of feeders should be multiplied by six to arrive at the daily total.

As mentioned on *pages 15* and *34*, it seems likely that hummingbirds use their color vision to find flowers, and that red flowers seem to be particularly favored as they are most easily visible against a background of green. For this reason, many hummingbird feeders are colored red. Studies undertaken to try to establish whether hummingbirds have a color preference have, however, reached very different conclusions. Even the most sophisticated studies have been unable to discriminate between real preferences and memory, and it may simply be that red, as it stands out, is recognized as a good source of food, and that, once a red 'flower' is found, hummingbirds simply become conditioned to feeding from it.

While the use of hummingbird feeders may be beneficial for the hummingbirds, and for people who wish to see them at close range, it may not necessarily be a good thing for flowering plants. Hummingbirds using feeders carry far less pollen than do other hummingbirds, suggesting that the flowers in the vicinity could possibly be losing one of their chief pollinators. Nevertheless, whether at a backyard feeder, at feeders in a remote forest reserve, or when naturally visiting wild flowers, hummingbirds are endlessly fascinating to watch. Although not all aspects of a hummingbird's extreme physical adaptations can be seen, their more obvious anatomical features and their incredible abilities in flight can all be appreciated at close range as they concentrate on doing what they must do to survive, often oblivious to the presence of a human admirer. Such encounters provide endless joy to those who are prepared to take the time to watch and wonder about the marvels of the natural world, and they also provide an ideal opportunity to appreciate a hummingbird's coloration, another fascinating adaptation of these remarkable birds.

ABOVE Feeders provide an ideal opportunity to watch hummingbirds at close quarters and to observe how far they are able to extend their tongue beyond the tip of their bill. This is a female **Anna's Hummingbird** *Calypte anna*. | Canada

BELOW The feeders at the Fundación Jocotoco **Buenaventura Reserve** in the province of El Oro, in southern Ecuador, attract good numbers of a range of hummingbirds: in this case mostly **Green-crowned Brilliants** *Heliodoxa jacula*, a few **Green Thorntails** *Discosura conversii* and a couple of **Violet-bellied Hummingbirds** *Amazilia julie*.

Color and Iridescence: Flashes of Brilliance

Perhaps what is most captivating about hummingbirds, aside from their amazing flying abilities, is their incredible colors and, in many species, stunning iridescent plumage. Seemingly, in the blink of an eye, a bird that at first appears dull black or green can be transformed, as if by magic, into a profusion of shimmering hues. And while this transformation is not magic, but instead simply the effect of waves of light interacting with feathers, the reality is arguably even more amazing.

Although hummingbird feathers are capable of presenting the full spectrum of colors, it is red, orange and yellow that are the least commonly encountered. When it comes to greens, turquoise, blues, purples and purple-reds, however, the hummingbird palette runs wild. All of these colors have clearly long been important to us in our appreciation of these little jewels, as demonstrated by the delightfully evocative English names given to many species. These names reflect not only their color but often a particular quality or manifestation of that color, as well as gemstones or metals, or include celestial terms (as summarized in the table *below*).

As stunningly beautiful as hummingbirds are, their fast movements and the physics that might allow only a glimpse of their iridescence often make them difficult to appreciate fully. A brief view of a hummingbird might enable us to get some idea of size and shape or flight action, but not so much of that all-important color. A dark shape, or perhaps a flash of white may be all that can be perceived if the hummingbird is simply zooming away into the distance. But sometimes these spectacular birds are obliging enough to return to feed at a flower, or perch on a twig, or visit a special feeder, giving the lucky viewer a chance to truly enjoy the bird. And in those sublime moments when they are caught by a ray of sunshine breaking through foliage at a precise angle to the viewer, the hummingbird is transformed into a creature beyond compare—as illustrated by the portfolio of images of aptly named species on the *following pages*.

Hummingbird names (note: the names of some species, *e.g.* **Coppery-headed Emerald**, use more than one term)

TYPE	TERMS
GEMSTONES* **53 names using**	amethyst, beryl, emerald, garnet, gem, ruby, sapphire, topaz, tourmaline
METALS **33 names using**	bronze, copper, gold, metal, steel
CELESTIAL TERMS **59 names using**	comet, lucifer, rainbow, star, sun
MANIFESTATIONS OF COLOR **33 names using**	brilliant, fire, glittering, glowing, scintillant, shining, spangled, sparkling, velvet
COLOR **109 names using**	azure, blue, brown(s), green(s), hyacinth, indigo, lazuline, magenta, orange, pink, purple, red(s), turquoise, violet

*Hummingbirds' gem-based names are due more to human wonder than to accurate mineralogy, as many gemstones occur in a range of colors. Sapphires are a good example, being generally regarded as blue, but in reality occur in any color except red (which would make it a ruby). Narrow color ranges are found in emeralds (always a bright blue-green), rubies (pure red to slightly purplish-red) and amethysts (reddish-purple to purple). Wide color ranges are found in, for example, tourmaline, which can be anything from blue-black to blue, red, yellow, green and pink, and beryl, which ranges from red to blue-green; the two can also be colorless.

The iridescent feathers on the back of the aptly named **Shining Sunbeam** *Aglaeactis cupripennis* contrast with the rather dull flight feathers. | Colombia

FACING PAGE **Red-tailed Comet** *Sappho sparganurus* | Bolivia

The true gems of the bird world

A total of 53 hummingbirds are named after precious or semi-precious gemstones, reflecting their scintillating iridescent plumage. Five are shown here: BELOW **Brazilian Ruby** *Clytolaema rubricauda* (*top left* | Brazil); **Emerald-bellied Puffleg** *Eriocnemis aline* (*top right* | Peru); **White-throated Mountain-gem** *Lampornis castaneoventris* (*bottom left* | Panama); **Amethyst-throated Sunangel** *Heliangelus amethysticollis* (*bottom right* | Ecuador); FACING PAGE **Sapphire-spangled Emerald** *Amazilia lactea* | Brazil.

Manifestations of metals

Thirty-three species of hummingbird have some reference to metals in their name, four of which are shown here: BELOW **Coppery-bellied Puffleg** *Eriocnemis cupreoventris* (*top left* | Colombia); **Golden-breasted Puffleg** *Eriocnemis mosquera* (*top right* | Ecuador); **Blue-throated Goldentail** *Amazilia eliciae* (*bottom* | Costa Rica); FACING PAGE **Coppery Metaltail** *Metallura theresiae* | Peru.

From the heavens

No fewer than 59 hummingbirds have been given English names that relate to celestial terms, reflecting their fantastic iridescent plumage—the words used most frequently being 'sun' and 'star'. A selection of the species that have been so named is shown here: BELOW **Rainbow Starfrontlet** *Coeligena iris* (*top* | Ecuador); **Blue-throated Starfrontlet** *Coeligena helianthea* (*bottom left* | Colombia); **Purple-throated Sunangel** *Heliangelus viola* (*bottom right* | Ecuador); FACING PAGE **Ecuadorian Hillstar** *Oreotrochilus chimborazo* | Ecuador.

Manifestations of color

Hummingbirds are frequently given wonderfully evocative names that relate to a manifestation of color or to their overall appearance, some examples of which are shown here: BELOW **Magnificent Hummingbird** *Eugenes fulgens* (*top left* | Costa Rica); **Snowcap** *Microchera albocoronata* (*top right* | Costa Rica); **Glittering Starfrontlet** *Coeligena orina* (*bottom left* | Colombia); **Shining Sunbeam** *Aglaeactis cupripennis* (*bottom right* | Colombia); FACING PAGE **Velvet-purple Coronet** *Boissonneaua jardini* | Ecuador.

Why be colorful?

The colors of a hummingbird's plumage, as with other birds, either help an individual to 'hide', by making it less conspicuous, or do the opposite and help it to show off in territorial and courtship displays. Furthermore, while some predatory birds have a cryptic coloration in order to be less easily detected by their prey, hummingbirds, being chiefly nectarivores (nectar-consumers), do not have the need for such subterfuge.

Most species of hummingbird are sexually dimorphic (see *pages 82–85*), the females typically having much duller plumage than the males as they need to be camouflaged when incubating their eggs or feeding young in order to avoid detection by predators. Many male hummingbirds, on the other hand, and particularly of those species that occur in open habitats, have bright, colorful plumage that makes them particularly conspicuous. It has been suggested that the amount of energy and effort that an individual male puts in to ensuring that his feathers are maintained in prime condition may signify his level of fitness. In such instances, a male hummingbird could be balancing the risks of drawing attention to himself (when defending a feeding territory or searching for a mate, for example) against the benefits of looking super-fit (or simply beautiful) and, therefore, extra special in the eyes of a female. In any case, male hummingbirds will never give away the location of a nest, because they usually have nothing to do with constructing the nest, incubating the eggs, or caring for the young, and at no time do they pay a visit. There is, however, another line of thought that, in some species at least, males may be brightly colored in order to attract the attention of predators and draw them away from the females, nests and young. While this may seem a very risky strategy, in evolutionary terms it may actually be effective as it increases the chances of the male perpetuating his genes.

There is some evidence to suggest that the color intensity of a hummingbirds iridescent feathering is influenced by diet. A high-protein diet has been shown to create brighter colors on the crown of male Anna's Hummingbirds, as well as a richer yellow-green on the tail feathers. This perhaps provides an additional indication that there are visual clues when showing off their colors in courtship and aggressive display as to which are the 'fittest' males. Although, to us, all male Anna's Hummingbirds may look pretty much the same, from a hummingbird's perspective they are likely to be recognizable as individuals, especially when taking into consideration the additional colors and hues that hummingbirds are able see that we cannot (as explained in the diagram on *page 62*).

Coloration may play a part also when a male is trying to maintain sole control of a patch of nectar-rich flowers—bold colors acting as a warning signal to other males and reducing the chances of an aggressive interaction, with all its inherent risks. As well as using color on its own, some hummingbirds show color patterns and areas of contrast in their plumage. A male Collared Inca, for example, has no bright colors but nonetheless is visually striking owing to an intensely black area surrounding a spotless white collar. On the whole, though, the males of forest-dwelling species are not particularly brightly colored, presumably because not much sunlight, an essential requirement of iridescent color, penetrates down into their dark and gloomy habitat.

The need for camouflage to reduce the risk of predation is another important aspect of hummingbird coloration. A subtly constructed pattern of contrasting streaks and spots can break up a bird's outline and help it to blend into its surroundings. An alternative is to rely on countershading, whereby light from above makes a bird's dark upperside appear paler and its pale underside (in shadow) look darker. This essentially reduces contrast and 'flattens out' the appearance of the three-dimensional bird, making it harder for a predator to spot.

It would seem logical that a hummingbird's bright, iridescent plumage has an extra impact compared with the 'ordinary' bright, pigment-based colors seen on many other birds. When a hummingbird really wants to show off, it can, by turning its head or spreading feathers in a particular way, 'turn on the lights' and send a strong directional signal that will be visible only from a certain angle (as explained in the diagram on *page 68*).

The plumage of 'typical' hummingbirds is more varied than that of the hermits, the males of many species showing extensive iridescence of red, orange, yellow, green, blue, purple or pink. These colors may be enhanced by various adornments, such as a fan-like gorget, a crest or long tail feathers. On the other hand, the plumage of hermits is predominantly a range of muted browns, grays and reds produced by pigments, with iridescence, if there is any, mostly limited to the upperparts—far less eye-catching than that of other hummingbirds. Most hermits are non-territorial and, as explained in the following chapter, *Breeding: Continuing the Line* (*page 87*), males display communally in what are known as leks, in which the color of the bill and open gape plays an important role in the display. Hermit leks are often in gloomy habitats where the lack of light makes a glowing plumage nearly impossible to achieve. The fact that the gape appears as a flash of color suggests that color must have a function, and leads to the conclusion that in hummingbirds, whether iridescent or not, color generally performs an important role in display.

As mentioned earlier, the finest coloration seems to play at least some role in attracting females, although exactly how a female makes her choice of mate is much

ABOVE **Collared Inca** *Coeligena torquata* | Ecuador; BELOW **Black-bellied Hummingbird** *Eupherusa nigriventris* | Costa Rica

more difficult to explain—and a topic that is still much debated—from the established 'finery = fitness' theories to the thought that perhaps beauty evolves simply for aesthetic reasons, for its own sake. It is, however, very unlikely that a female hummingbird chooses a male simply because he looks good, but rather that she chooses him because he is giving a message about his fitness to father a strong brood of chicks. A fascinating line of thought that springs from this is that, if females prefer these colorful males, then the evolution of such males puts them at greater risk of predation: in such cases it could be argued that evolution is reducing the chances of an individual contributing to a future gene pool, which would seem paradoxical. This does, however, provide the basis for the so-called 'honest signaling theory', which suggests that because certain traits, such as a colorful plumage, have a high maintenance cost and present a higher predation risk there is no real opportunity for individuals to 'cheat the system'. Only the highest-quality individual males can afford the risks associated with this type of signaling, and prospective mates or rivals can easily see a visual clue to the fitness of these 'honestly displaying' individuals.

So the key question remains, why be colorful at all? Is it merely to look beautiful and to be appreciated, or is it, as most scientists believe, to relay some sort of message about fitness and willingness to risk the 'costs'? In other words, does coloration confer some competitive advantage with risks attached, or are colors merely arbitrary developments that offer little or no gain in an evolutionary sense? When considering this last point it is important to take into account the fact that sexual selection is the strongest driver of evolutionary traits in the natural world.

How does color work?

When contemplating how color works, it is important to put anthropocentric views to one side and to look at light purely as it exists, recognizing that color is merely an interpretation of wavelengths within the limitations of what we can actually see. Hummingbirds, like many animals, and indeed plants, are 'simply' taking advantage of the laws of physics relating to light. But the more one delves into the subject, the more one begins to realize that this simplicity is in fact quite complex and requires an astounding set of circumstances to make it happen.

Light is a form of energy (electromagnetic radiation) that acts as both waves and particles (photons), which is just a small portion of the electromagnetic wavelengths that range from very low-frequency radio waves, through sunburn-inducing ultra-violet, to extremely high-frequency gamma rays (see the figure *below*). Light (which on Earth means predominantly sunlight) is a beam of photons in the visible portion (between ultraviolet and infrared) of the electromagnetic spectrum. Sunlight appears 'white' unless refracted by some means—in which case the light splits into the familiar rainbow spectrum. The color that is actually seen by an observer depends on which wavelength is reflected back.

It is these waves that are fundamental to color, as our eyes perceive different wavelengths (measured in nanometers (nm)) of light and our brains interpret them as different colors. For example, a typical rainbow is created when sunlight passes through water droplets in front of a viewer at a precise angle of 42°. Sunlight is bent (refracted) when it enters the almost round, clear water droplet before being reflected by the back of the droplet and refracted once again upon exiting the droplet; whichever wavelength leaves the droplet is perceived as a color. The same principle applies when

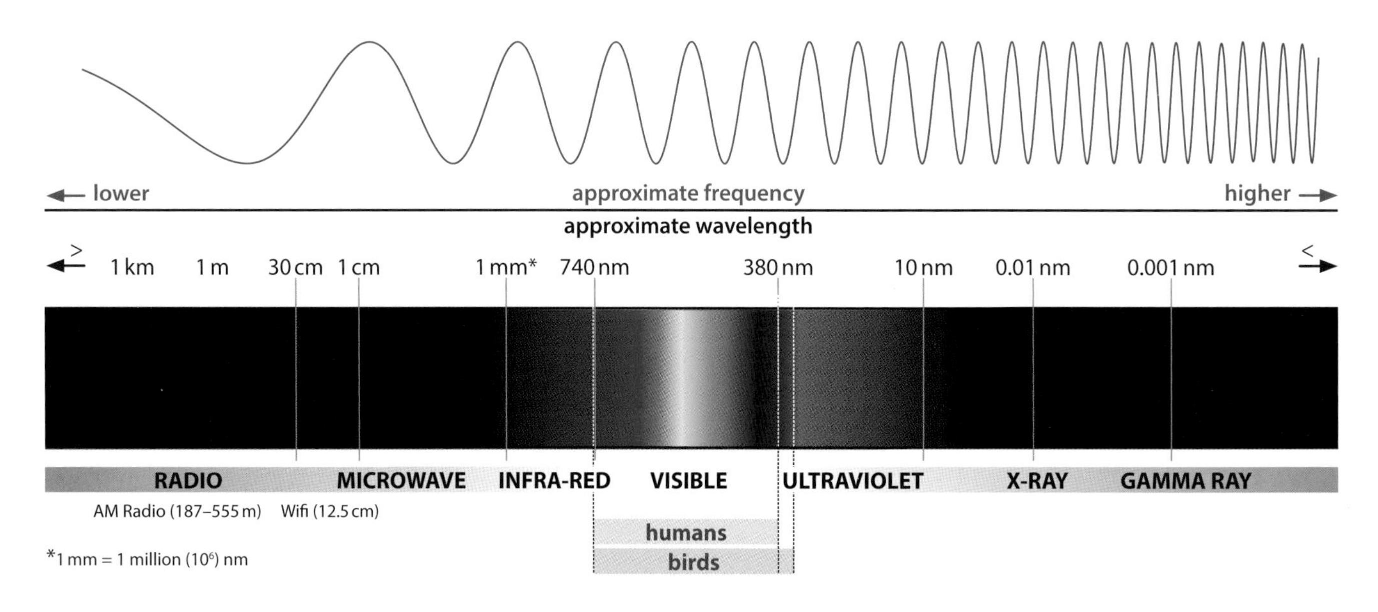

The electromagnetic spectrum showing the categorization of waves and the range of visible wavelengths for humans and birds.

Bright, often iridescent, colors are a prominent trait of hummingbird adornments: **Wine-throated Hummingbird** *Atthis ellioti* (*top left* | Guatemala), **Rufous-crested Coquette** *Lophornis delattrei* (*top right* | Peru) and **Long-tailed Sylph** *Aglaiocercus kingii* (*bottom right* | Ecuador). On hermits, vivid colors are confined to the bill and gape: **Long-tailed Hermit** *Phaethornis superciliosus* (*bottom left* | Brazil).

Color		Wavelength range (nm)
red		750–625
orange		625–590
yellow		590–565
green		565–500
cyan		500–485
blue		485–450
violet		450–380
ultraviolet		380–300 (birds, not humans)

Visible light wavelengths

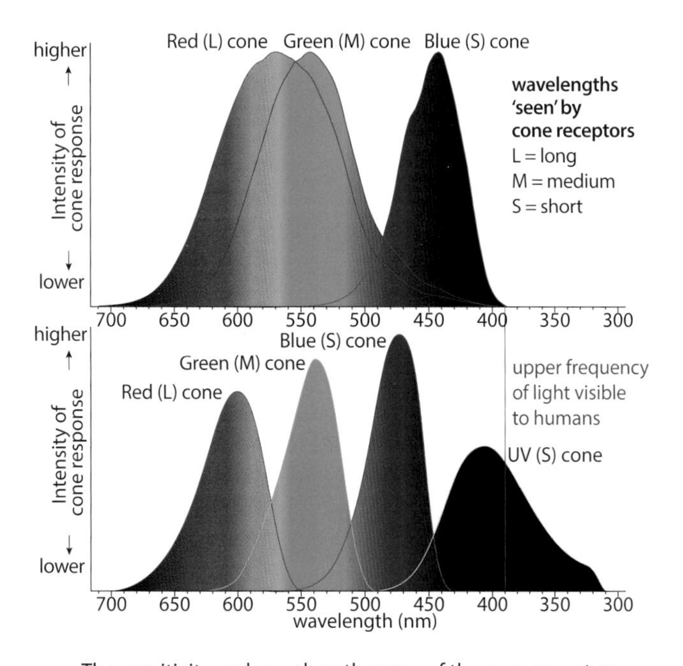

The sensitivity and wavelength range of the cone receptors in a typical human eye (*top*) compared with that of a typical hummingbird (*bottom*).

viewing oil on water, or a soap bubble, although in this case, the color seen is a function of the thickness of the surface film, which affects the speed at which the light travels through and is refracted—the different thicknesses causing different wavelengths to reflect back, producing a distorted rainbow effect (see the box on *Interference colors* on *page 66*).

What wavelengths (and therefore colors) can be interpreted is dependent on the receptors in an animal's eyes. Human eyes have three types of cone receptor (red, green and blue), whereas hummingbirds' eyes (as those of most birds) have four (red, green, blue and a fourth that is sensitive to some ultraviolet light). The visible light spectrum available to humans (as seen in a rainbow) has wavelengths between approximately 730 nm (red) and 390 nm (violet). The extra cone that a hummingbird possesses allows it to perceive wavelengths in the near ultraviolet zone (between 400 and 300 nm), and, as a result, it can see a broader range of colors than us, including nonspectral colors (see the graphs *left*). This ability of hummingbirds to see ultraviolet colors has been proven by testing their capacity to discriminate colors based on colored feeders containing either a sugar-rich solution or pure water. The results of these studies showed that hummingbirds could discriminate colors, including ultraviolet-influenced nonspectral colors such as purple, ultraviolet+green, ultraviolet+red and ultraviolet+yellow. While the ultraviolet+green feeders (sugar) and the green feeders (water) looked identical to the researchers, the hummingbirds almost unerringly showed a preference for the sugar-rich ultraviolet+green feeders. This suggests that the role of color and iridescence in hummingbirds may be even more important than currently thought and very likely serves a key role in both display strategies and foraging efficiency.

As well as the color-creating process of refraction (which is explained later in this chapter), there are two other aspects of light that also play fundamental roles in how colors are seen: absorption and reflection.

Black, brown, pink and white

Some materials, especially those that are matt black, absorb almost all light. This means that light hitting that surface is not reflected in any great quantity but is instead converted into heat energy (which is why, in sunlight, black surfaces feel warmer than white surfaces). Conversely, some materials, typically those that are opaque or highly polished, absorb virtually no light but instead reflect it. For example, on a sunny day, an open window or a car roof reflecting the sun can generate an intensely piercing, star-like spot of silver-white light that can easily be seen with the naked eye from miles away. Because so little light is absorbed by the glass or metal, the full spectrum of sunlight is reflected as a white beam (this was the basis of signaling by heliograph long ago). Many other colors that we commonly differentiate are a result of the brightness of the light reaching our cones. For example, black does not activate any cones at all; brown activates the red cone partially and the green cone slightly; and flesh (pink) tones are a result of the red cone being fully activated, the green cone a little less and the blue cone slightly.

Hummingbirds presumably have a similar method of color perception to us, although their additional UV-sensitive cone opens up an even wider range of both spectral and nonspectral color options. For example, the reflectance of the Broad-tailed Hummingbird's gorget starts at approximately 300 nm (or possibly lower). This is outside our perception range, but to another hummingbird will appear as colors about which we can only speculate. Also, any activation of the UV cone in combination with any of the other three cones in a hummingbird's eyes will result in a greater range of colors that humans cannot perceive—UV-Red, UV-Green, and UV-Blue.

Nonspectral colors—the magenta conundrum

Look at a rainbow and ask yourself the question "Where is the pink?" The answer is that this 'color' is not actually there! This is because pink or, to be more accurate, magenta, does not exist in the spectrum as a spectral color (*i.e.* one with a wavelength as found in the rainbow) but instead exists as a nonspectral color, which is one perceived as a result of the simultaneous stimulation of cone types that are sensitive to different, even widely separated, light wavelengths.

When light is received by the various cones in an animal's eye, it is processed depending on the light wavelengths that the cone is stimulated by—with a higher or lower intensity of stimulation determined by the actual wavelength involved. If two cones are activated by the same beam of light, then the brain processes the cumulative effect of the wavelengths received. The graph *opposite* (*page 62*) shows that humans do not respond to 'darker' reds with as high an intensity as we do to more orange-reds, and that we are best at differentiating the subtleties of a range of yellow to green colors. This is because both our red-sensitive and green-sensitive cones overlap considerably in the wavelengths they receive, and in the intensity of cone response that those wavelengths create.

If both our red-sensitive and blue-sensitive cones are stimulated equally, we might expect the color mix perception to follow this same 'average' mix pattern and be midway between the two spectral wavelengths (*i.e.* green). However, we already have a cone that is sensitive to green, and therefore a mix of red and blue creating green makes no sense to our brain! So, instead, we perceive these mixed wavelengths of red and blue light as magenta to purple, colors that do not exist within the spectrum of light or as a result of spectral colors affected by the intensity of light, such as pink and brown. Put simplistically, our brain 'invents' these colors and the magenta and purples we see on many hummingbirds are an illusion (so-called 'chimerical' colors—as explained graphically to the *right*).

Hummingbirds, however, appear to be most sensitive to, and therefore respond to, relatively narrow ranges of red-orange, green and blue and, to a lesser extent, their ultraviolet range. This suggests that these limited colors are of importance, and that the ultraviolet component of their color vision is also significant.

The logical next question is that of how hummingbirds take advantage of the physics of light—and in particular the effects of absorption, refraction and reflection.

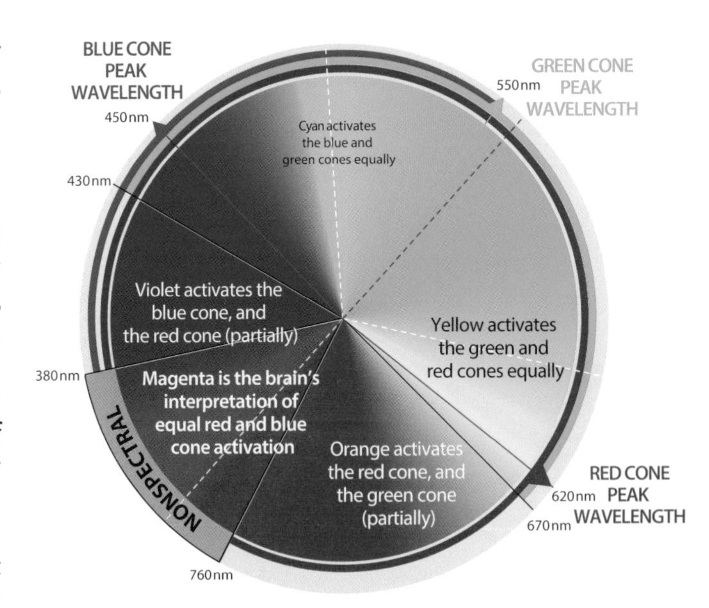

If the same 'equal mix rule' seen in other spectral colors was followed, then an equal mix of red and blue light should be seen as green ...

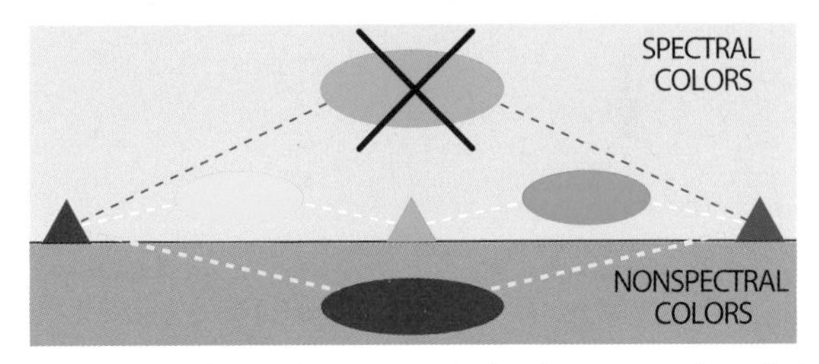

... instead, the human brain 'invents' colors (magenta and purples) between the red and blue wavelengths.

How magenta is created in our brain
The midway points of equal stimulation of two cones (white dotted lines) give red + green = yellow, and green + blue = cyan. On that basis (magenta dotted line) red + blue = green; instead, we see magenta (yellow dotted line).

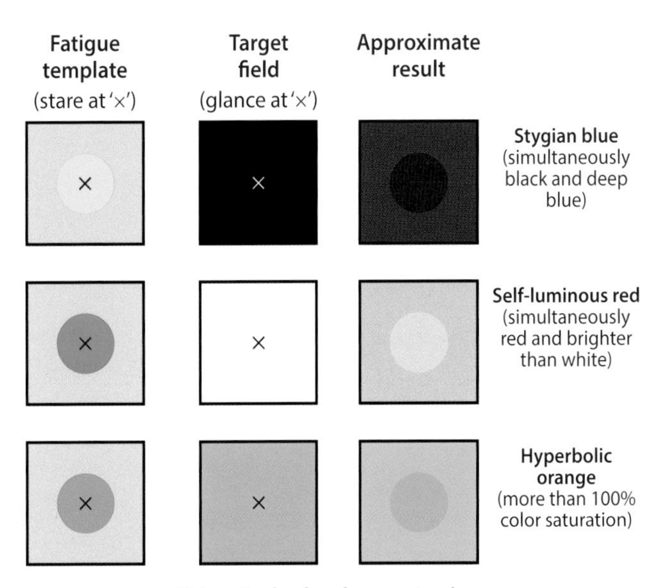

Chimerical color demonstration
Staring at the '×' in any of the fatigue fields above for 20–60 seconds will fatigue your color receptors; glancing at the associated target field '×' should result in some color-based strangeness!

The body and tail feathers of many hummingbirds are brightly colored, whereas the flight feathers are dark and dull, reflecting the presence of strengthening melanin pigment—as shown by this **Buff-winged Starfrontlet** *Coeligena lutetiae.* | Ecuador

How does feather color work?

Birds' feathers utilize the absorptive, refractive and reflective qualities of light to create colors in one of two ways: either through pigmentation within the feather or through a feather structure that has evolved to create certain colors. Although these two methods work in isolation, they can also work in combination.

Pigmentation

On most species of bird, the colors that we see are produced by pigments, deposited on the feather surface, which absorb and reflect particular wavelengths of light as a single color. Feather pigments fall into three main groups:

Carotenoids—which absorb shorter wavelengths while transmitting and reflecting longer wavelengths, producing bright yellow and orange. These pigments are found in plants and are obtained by birds either directly, by eating the plant, or indirectly, by eating something else (such as a caterpillar) that has eaten the plant.

Porphyrins—which produce pink, brown, red and green hues. Porphyrins are created from specific amino acids and fluoresce bright red in ultraviolet light, suggesting that they may have evolved to stimulate birds' UV-sensitive cones.

Melanins—these pigments have a high absorbance of visible light, the highest absorbance being of shorter wavelengths resulting in black or golden to rusty-red coloration. Two types of melanin are involved, eumelanin and phaeomelanin, respectively, and these can mix to produce intermediate colors, such as brown, and combine with carotenoids to produce shades of olive-green. Melanins are produced by the oxidation of the amino acid tyrosine in special cells found in feathers and skin.

Melanins are particularly important in birds' feathers, since not only do they produce colors ranging from intense black to reddish-browns and pale yellows but they also add strength and stiffness to the feather structure. For this reason, they are frequently present in flight feathers, or at least at the tips of the longest primary feathers, which have to withstand a lot of wear and tear. Hummingbirds possess both the blackish and rufous basic pigmented color types produced by melanins, and, since flight is so crucial to their lifestyle, the wing feathers are invariably melanin-rich—dark-pigmented and dull.

Feather structure

Notably absent from the list of colors produced by pigments is blue, a color very commonly found in birds. Blue is very difficult to achieve through pigmentation, but much easier to produce as a result of a feather's structure 'directing' the way light is absorbed, refracted and reflected. This is because blue light has a very short wavelength and so is more easily reflected than colors with longer wavelengths. The blue sky we see during the day (when the sun is at a steeper angle to the Earth) is caused by individual nitrogen and oxygen molecules scattering this easily reflected blue light, which effectively blocks out the other spectral colors. Conversely, in the evening, when the sun's angle is lower and light has to travel farther through the atmosphere to reach us, the wavelengths that are predominantly scattered are the longer reds and oranges.

The blue feather colors of birds make use of this scattering, reflective quality of blue light—not in a random way (as in the sky) but rather by having a feather structure that is designed to interfere with the light in an organized manner. This is known as 'structural coloration'. Blue feathers typically have a dark, melanin-rich layer (that absorbs the longer wavelengths), which, either as an arrangement of pure melanin or with small air cavities and/or keratin particles on its surface, reflects and scatters blue light.

White feathers also appear the way they do due to their structure. They reflect nearly all light by virtue of a complete absence of melanin combined with a microscopic surface topography that resembles cut glass or snow. The white feathers of some species also contain air pockets that increase the combined total reflection of all colors of visible light, making them appear even brighter. White feathers can be shown to lack any pigment as they become transparent when immersed in an effective light-blocking substance such as balsam.

Colors that result from structural effects of feathers, such as the intense blues displayed by many kingfishers worldwide, are termed semi-iridescent insofar as (unlike the perfect alignment of iridescent feather structures) the microscopic structures that create the blue color are imperfectly aligned, causing a narrower range of iridescence that can be seen from a wider range of angles. The intense brightness can be explained by two sets of reflected waves being in phase and hence effectively doubling up (as explained in the diagram on *page 66*).

The resemblance of some hummingbirds to sunbirds of Africa and Asia is mentioned in *An Introduction to Hummingbirds* (see *page 9*), but sunbirds, bright and beautiful as they are, usually have colors that are more constant, being less dependent upon the exact angle of view. The basic reds, greens and yellows of the sunbirds are for the most part visible all of the time by virtue of bright red or yellow carotenoid pigmentation (which hummingbirds lack), but sunbirds also have iridescent feathers more akin to the semi-iridescent feathers of a kingfisher or the head of a male Mallard duck.

Although structural coloration has been broadly understood since 1665, when Robert Hooke observed that a peacock feather was iridescent in air but not under water, drawing the conclusion that pigments could not be responsible, the mechanisms of iridescence were not fully understood until the turn of the 19th century. It is only very recently, however, through the advances of technology, that we are really beginning to understand the complexities and incredible accuracy of the structures involved in order for iridescence to occur.

Iridescence

The particular shades of lustrous green on the body of various 'emerald' hummingbirds is pretty much the same on each and every individual of a given species. Similarly, the fierce shot of red on a Crimson Topaz (see *page 72*), seen full-on in good light, is always the same shade of pure crimson. In many hummingbirds, however, just parts of the plumage, such as a shiny gorget, are iridescent, looking bright red when viewed head-on, bluish when seen from slightly to one side, but dull black from a greater angle.

True iridescence is a combination of how light interacts with complex biological nanostructures that produce thin films or so-called 'diffraction gratings' in a constant fashion. In hummingbirds this is no random event but rather a result of astoundingly precise feather structures that are perfectly aligned and cause the light

Different ways in which colors are created in feathers (the arrowed line indicates the feather's cross section from keratin to basal layer).

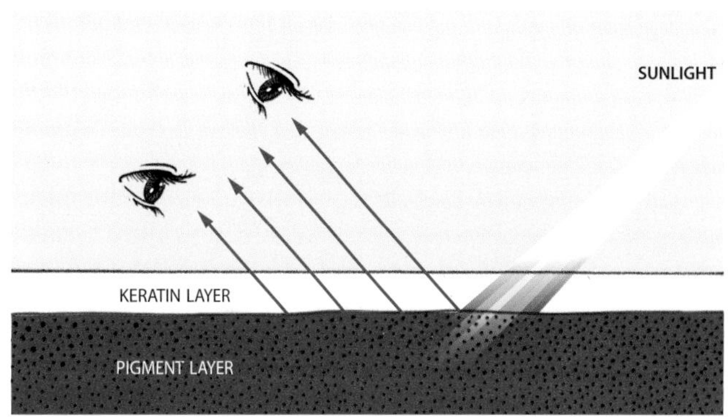

Pigmentation: pigment granules in a feather selectively absorb and reflect particular wavelengths. For example, carotenoid pigments absorb shorter wavelengths (such as blue) and reflect longer wavelengths (such as red). Light is reflected only as a single color and this color is constant no matter the viewing angle.

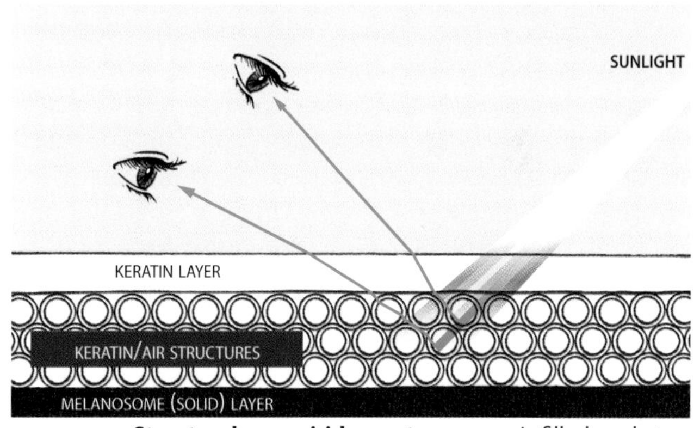

Structural—non-iridescent: spongy, air-filled pockets in a feather selectively absorb particular wavelengths and reflect/refract others, creating a scatter pattern of a single wavelength, and hence a constant color, when viewed from any angle.

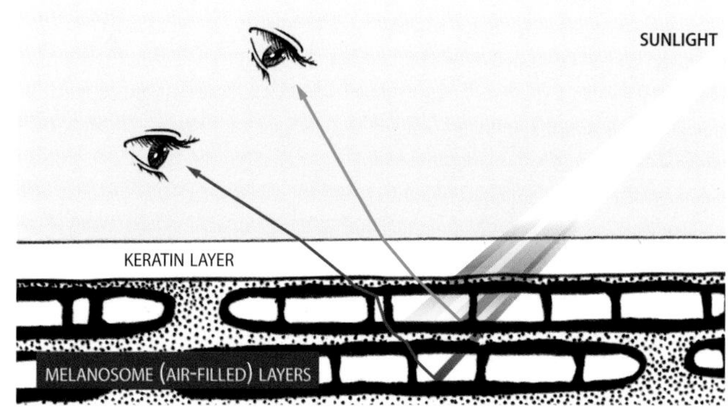

Structural—iridescent: spongy, air-filled pockets in the melanosome layer of a feather selectively absorb some wavelengths and reflect/refract others. The complexities of the structures and air pockets (different thicknesses of melanin and air) mean that waves can either be reflected directly or be 'bounced around' by being refracted multiple times within the structure before exiting. If the wavelengths are in phase they add together to create a brighter, more intense version of the color; if they are out of phase they cancel each other out and no light is returned. At some angles a particular light wavelength may simply be reflected back on itself and be unable to exit the structure. The diagram above is highly simplified; in reality, multiple wavelengths interact both constructively and destructively to create different colors with a range of shimmering brightnesses and intensities (see *page 70*).

to refract and reflect in a variety of ways to produce waves that are termed interference colors (as seen randomly in, for example, a soap bubble—see box *below*). This accuracy of structure and the resultant coloration are seemingly critical from an evolutionary perspective, as they are highly significant in terms of breeding success.

In the iridescent feather of a hummingbird, the basal two-thirds or so are unexceptional and look like a typical feather, but the tip (the part visible when the feathers are smoothed down) is where the highly specialized structures are located. Birds' feathers are made of keratin, which is the same substance as human hair and nails. The feather barbs that are attached to the main shaft (or rachis) have, in turn, tiny filaments attached to them called barbules. Each barbule of an iridescent hummingbird feather has a reflective surface layer of keratin seated above a mosaic of microscopic packets of melanin called melanosomes. In the iridescent feathers of most birds, these melanosomes are log-shaped and solid. In hummingbirds, however, they are typically pancake-shaped and filled with tiny air pockets, although in some feather tracts of some species there may be solid melanosomes, or a mixture of solid and air-filled. Astonishingly, in a hummingbird's feather there are around 1,500 of these 'bubbles' per square centimeter (about 10,000 per square inch). Depending on the species, feather keratin comes in a range of thicknesses; melanosomes vary slightly in shape, contain different air-bubble sizes and are stacked from seven to fifteen deep. This range of factors gives rise to a wide range of surface complexity and, as a result, the incredible variety of hummingbird iridescent colors that can be seen (see *pages 70–71*).

Light passing through the keratin, melanin and air will bend at slightly different angles, with the thickness of the granule and the depth of the air bubble each affecting the color of the reflected light. Thicker melanin with thinner air bubbles means that light with longer wavelengths (towards the red end of the spectrum) is reflected; thinner melanin and wider air bubbles means that light with shorter wavelengths (towards blue and violet) is reflected. This can happen, however, only if the reflected light is received by a viewer at the optimal narrow angle, as determined by the feather structure. Slight differences in angle will change the color seen, or its intensity or brightness, and, if viewed at the 'wrong' (or a suboptimal) angle, light within the melanosomes is all but fully absorbed, or undergoes destructive interference such that virtually no light is reflected and

Interference colors—the soap bubble example

When light waves meet the thin film of a bubble, light is reflected from both the outer and the inner surfaces. However, this means that the light reflecting off the inner surface has slightly farther to travel than that reflecting from the outer surface. Different light wavelengths will behave differently depending on the distance they travel through the film, which itself is dependent on the angle at which the lightwaves hit the film and the thickness of the film. If the distance traveled through the film is equal to, or is a whole number multiple of, a particular wavelength of light, then that light will reflect from both surfaces in such a way that the waves' peaks and troughs align perfectly. These waves are described as being **in phase**, and the result is an increase in color intensity and brightness known as constructive interference.

Conversely, if the distance traveled is half the width (or odd number multiple of) the wavelength, then the two waves' peaks and troughs will be perfectly aligned, effectively canceling one another out. These waves are termed as being **out of phase** and no color is visible (destructive interference).

In the cases where the film thickness means that reflected waves are neither perfectly in or out of phase, the results range from brightish to dullish colors depending on how much the waves are in or out of phase.

When all wavelengths are taken into account, then for any given thickness of film some wavelengths will be fully in phase, others fully out of phase and the remainder somewhere in between, mixing and matching to produce the familiar iridescence we see in bubbles. This is the essence of iridescence in hummingbirds' feathers, as the physics of interference colors is what hummingbirds have evolved to take advantage of via their complex yet precise feather nanostructures.

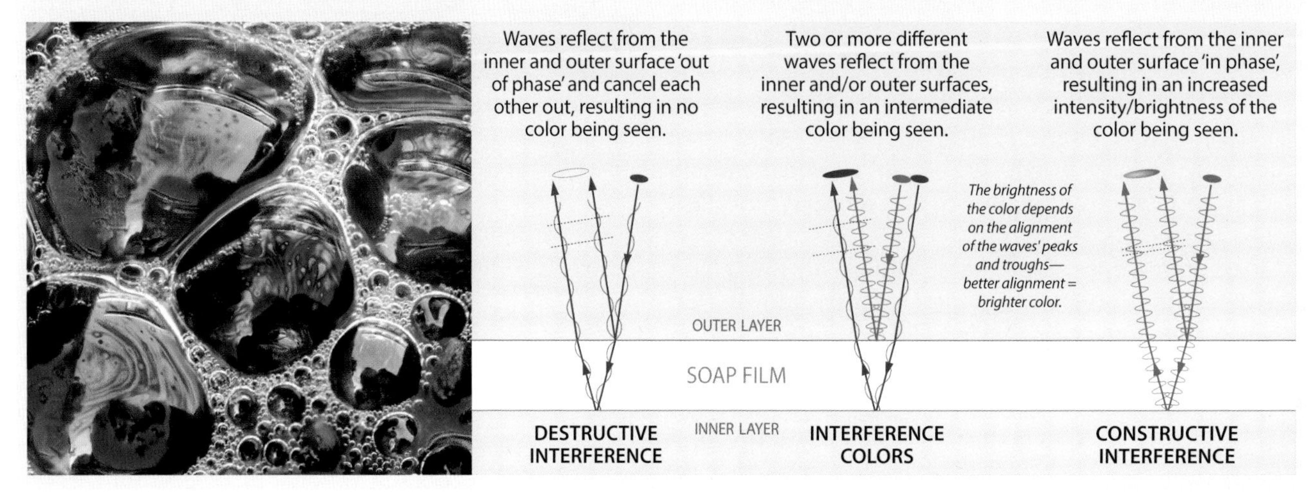

The constancy of coloration among individuals of the same species, such as these **Green Violet-ears** *Colibri thalassinus* (*above* | Costa Rica) and **Fiery-throated Hummingbirds** *Panterpe insignis* (*below* | Costa Rica), despite the effect of iridescence, is quite remarkable.

Iridescence

The structures behind the color

Hummingbird feather
Feathers are made of keratin and consist of a central shaft to which are attached many barbs, which in turn have barbules attached to them.

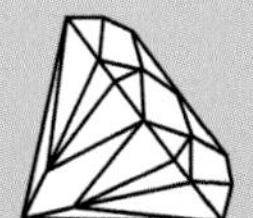

Iridescent hummingbird feathers look very shiny: their refraction index is just over 2 (for comparison, the refraction index of a diamond is 2.4).

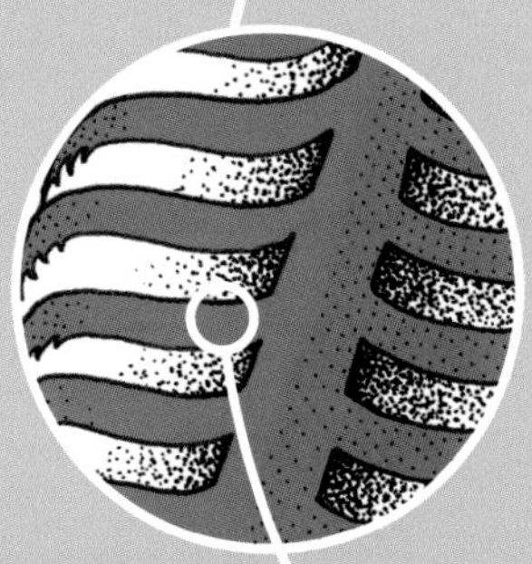

Feather barb with barbules
In many hummingbirds these barbules are twisted such that they enhance the role of the melanosomes in producing iridescent colors.

Melanosomes in layers
In Anna's Hummingbird the platelets are layered up to 17 deep—the air pockets within 'bend' and reflect the light that gives off the iridescent colors we see.

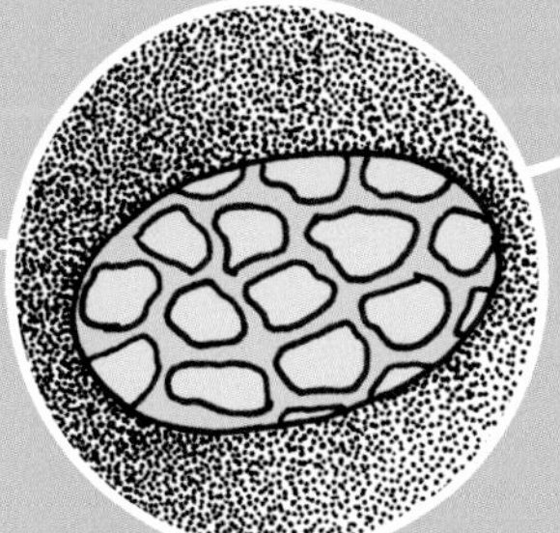

Barbule with melanosomes
Beneath a translucent layer of keratin the barbules of an iridescent feather have layers of melanosomes.

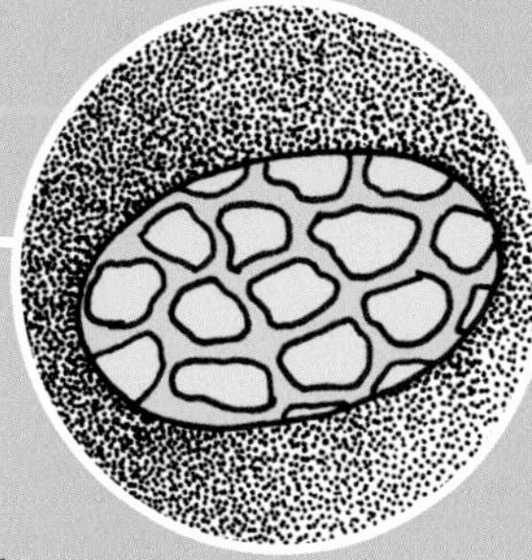

Melanosome
Pancake-shaped melanin platelets about 100 nm thick that contains pockets of air (viewed from above).

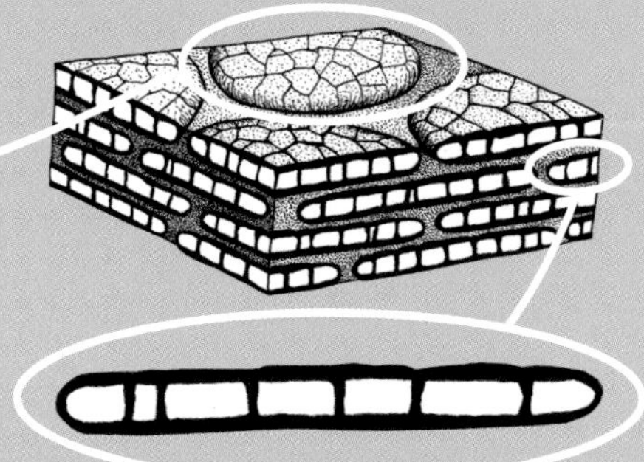

Melanosome cross-section
The thickness of the melanin and the shape and size of the air pockets is variable. This variation is key to the absorption and the refraction/reflection of different wavelengths of light which produces the iridescence we see.

so, to the viewer, the feather appears very dark or black (see *below*). In many iridescent hummingbirds, as well as these complex, yet accurately arranged melanosomes, the barbules are flattened and twisted to create the maximum effect for the viewer (whether that is another bird or, incidentally, any watching human).

Hummingbirds—a rainbow of iridescence …

While observing a hummingbird, at any given moment we might see a brilliant cap, a vivid throat patch, perhaps a shiny back, or maybe a brightly colored rump. Viewing angle is fundamental to the colors we (or indeed another hummingbird or predator) perceive. While a white patch may be seen from any angle of view, a shiny back or bright rump is most likely to be visible to us only when presented within a certain angle of view depending on the nature of the structural reflection. The full impact of an iridescent cap or throat patch will be apparent only when seen from the optimal narrow angle.

Given that sunlight is a single-point, unidirectional light source, it is, by the laws of physics, impossible to see all of these colors at the same moment, and much of the body of a hummingbird often simply looks 'dark'. In most instances, hummingbirds generally reveal their visual riches just one or two at a time. A Ruby-topaz Hummingbird appears as little more than a black-brown shape, darting from bloom to bloom high above us in a treetop. Even when lower down, viewed at a shallower angle, it can seem merely dull and dark, perhaps showing a brighter tail. With the light reflected at the optimal angle, however, there will be a glimpse of a ruby-red cap, before the bird turns face-on and, for a brief and brilliant second or two, we may be treated to a flash of its glistening deep golden-yellow throat (see *page 70*). This presents great difficulties for an artist, who is torn between giving a realistic impression of the bird and wanting to show the full range of colors that could, at any moment in time, be seen. And, of course, it causes similar challenges for photographers, who, without multiple simultaneous sources of light, have no possibility of capturing all the colors in a single shot.

The color red is clearly important to hummingbirds as it often seemingly influences their choice of

[continued on *page 78*]

These photos of an **Anna's Hummingbird** *Calypte anna* show the effect of iridescence as the bird turns its head. | Canada

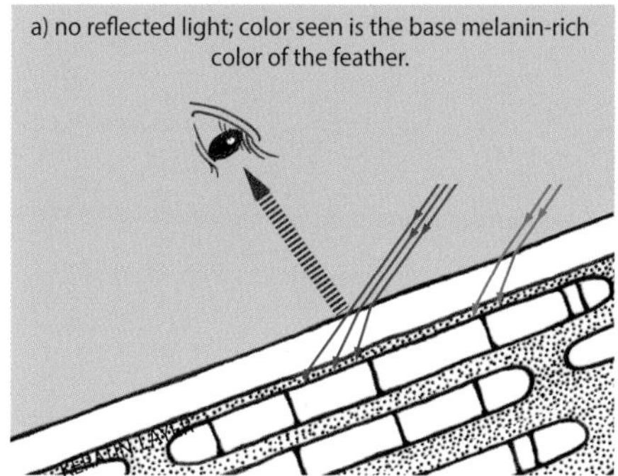

a) no reflected light; color seen is the base melanin-rich color of the feather.

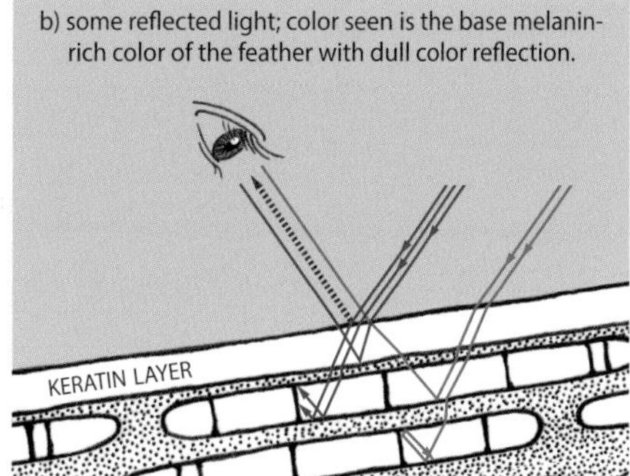

b) some reflected light; color seen is the base melanin-rich color of the feather with dull color reflection.

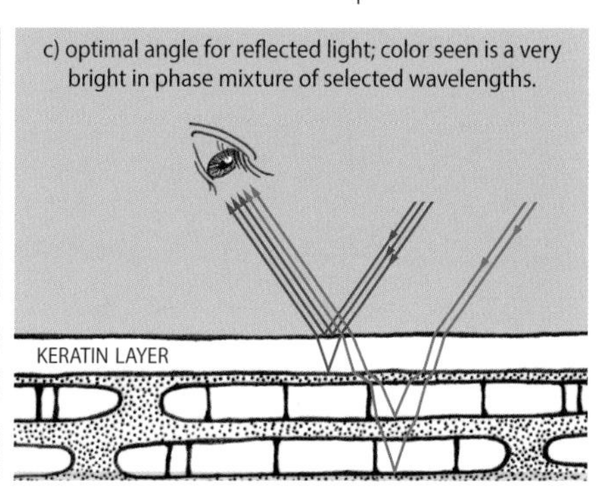

c) optimal angle for reflected light; color seen is a very bright in phase mixture of selected wavelengths.

Highly simplified illustration showing the effect of barbule angle on reflected light coming from the same direction
In Anna's Hummingbird, peak reflectance is in the red wavelength range (at approximately 670 nm), with a lesser peak in the blue wavelength range (480 nm)—the two combining to give a magenta iridescence (all other wavelengths are absorbed in the topmost layers of the barbule). In diagram a) the feather angle means that light does not reach any air pockets and is absorbed, such that the viewer sees only the melanin-rich base color of the feather. In diagram b) a slight change of feather angle means that some light reaches the air pockets, is 'bounced' around and a little is reflected, such that the viewer sees a magenta tint to the melanin-rich feather base color. In diagram c) the feather is at the optimal angle to allow the two wavelengths to be in perfect alignment with different air pockets and multiple waves are reflected in phase, such that the viewer sees a bright, intense mix of red and blue light (appearing as magenta) that obliterates the dark base color of the feather.

A rainbow of iridescence

Ruby-topaz Hummingbird *Chrysolampis mosquitus (above left* | Trinidad); **Orange-throated Sunangel** *Heliangelus mavors (above right* | Venezuela); **Magnificent Hummingbird** *Eugenes fulgens (below left* | Costa Rica); **Sapphire-throated Hummingbird** *Amazilia coeruleogularis (below right* | Panama)

Black-backed Thornbill *Ramphomicron dorsale* (*above left* | Colombia); **Crowned Woodnymph** *Thalurania colombica* (*above right* | Costa Rica); **Velvet-purple Coronet** *Boissonneaua jardini* (*below left* | Colombia); **Purple-throated Woodstar** *Calliphlox mitchellii* (*below right* | Ecuador)

Multiple colors

Hummingbirds exhibit the full range of colors, from reds, through greens to purples and purple-blues, often with patterning in discrete patches, as shown in the examples here and on the *following pages*:

BELOW **Crimson Topaz** *Topaza pella* | Guyana; FACING PAGE **Blue-tufted Starthroat** *Heliomaster furcifer* | Argentina

ABOVE **Royal Sunangel** *Heliangelus regalis* (*left* | Peru); **Gorgeted Sunangel** *Heliangelus strophianus* (*right* | Colombia);
BELOW **Green Hermit** *Phaethornis guy* | Costa Rica; FACING PAGE **Green Violet-ear** *Colibri thalassinus* | Panama

ABOVE **Violet-bellied Hummingbird** *Amazilia julie* (*left* | Colombia); **Festive Coquette** *Lophornis chalybeus* (*right* | Brazil); BELOW **Glowing Puffleg** *Eriocnemis vestita* | Colombia; FACING PAGE **Crowned Woodnymph** *Thalurania colombica* | Costa Rica

foodplant. But it is not such a common color in the birds themselves and only three have common names that start with 'red'—one describing the plumage, Red-tailed Comet (see *page 48*), and two that describe the bill, namely Red-billed Emerald and Red-billed Streamertail. There are, however, two strikingly red hummingbirds, the names of which reflect their particularly eye-catching hues—Crimson Topaz and Fiery Topaz. Other species are 'reddish', including one that is so-named—the Reddish Hermit—although in reality this bird is actually rufous-brown. Variations on the theme of 'red' also feature in other hummingbirds' names, notably ruby, which refers to the iridescent throat patch of the Ruby-throated Hummingbird and Brazilian Ruby, and to the tail in the case of the Ruby-topaz Hummingbird.

These scintillating reds merge into purples, which in turn blend into purple-blues, violets or a 'truer' blue. Some species are almost wholly purple or purple-blue (such as the Royal Sunangel), and the word 'purple' is used in the English names given to 11 hummingbirds, with 'blue' featuring in 14. Indeed, purples and associated purple-reds are among the staples in hummingbirds' common names. However, yet more species still are essentially 'green', albeit we can see in their plumages a wonderful range of greens. This rich assortment of greens may be correlated with green being a color towards which humans have great sensitivity (more so than hummingbirds), and hence we can distinguish a wider range of green–yellow hues. Twenty-three species are 'green' or 'greenish' and two 'olive' by name, others 'bronze' or 'bronzy', while no fewer than 28 are 'emeralds'. These greens are many and varied, from shimmering yellow-greens and golden-greens to the intense, acid greens of the Sparkling Violet-ear and Long-tailed Sylph. Often the base of a feather is dull brown or matt green, with a loose or fragmented crescent across the tip that shines iridescent golden-yellow, giving a glistening, shimmering effect, thus adding an extra dimension above and beyond a simple, uniform color.

... and black and white as well

Black, too, is a prominent 'color' among hummingbirds and 19 species have black in their simple yet descriptive names—Black-tailed Trainbearer and Black-throated Mango, for example. Wings and tails tend to look black, but are, in reality, brown-black; this relates to the presence of melanin, which, as explained earlier, provides added strength (see *page 64*). More genuinely black patches appear on cheeks and chins, under the belly or beneath the tail, sometimes in rounded spots on a patch of brighter hue, or in a 'T' shape on the spread tail. A few species, such as the Black Jacobin, are much more extensively black. Black on hummingbirds frequently merges into an area of iridescent red or green, such that the border between one color and the other is hard to define—the area shimmering and changing shape as a bird turns, which may well be another trick in the hummingbird's array of display behaviors.

Hummingbirds also use white feathering to great effect. Some species have patches of simple, pure white, often reflected in their common names (29 species): White-bellied Woodstar, White-throated Hummingbird, White-necked Jacobin and White-tailed Starfrontlet, for example, are clear and descriptive. White may be in discrete patches or fluffy tufts on the thighs, on the hindneck, beside the throat, as a cap seemingly tipped forward on the forehead, or as a tiny fleck behind the eye, under the tail, on the rump, or more extensively on the belly. The females of several species of mango have white on the breast and belly that is split by a ragged central streak of black (see *page 130*).

Some species of hummingbird have white on the tail, which tends to be particularly obvious when they are in flight. The Black Jacobin, for instance, has a striking white tail with a triangle of black extending beneath it from the belly, and a fine black central line and tip. A different pattern is shown by both the White-tailed Hillstar and the very similar Rufous-gaped Hillstar, which have a broad band of white down each side of the tail, with black in the middle and on the outer edge. Quite why such varied and complex patterns of white have evolved is hard to explain, but the white itself is surely a successful means of communication in a dark forest environment. The white on the neck, tail and belly of a White-necked Jacobin hovering above the ground along a forest trail is extremely eye-catching, as is the tail of a White-tailed Sabrewing against a background of green leaves and deep forest gloom in the mountains of northeast Venezuela or on Tobago. The white is flashed during courtship displays, and may be a warning in territorial defense, but, like all such marks on birds, it must confer some degree of risk, too, as it may also attract the attention of predators.

What does it all mean?

What sense can humans make of this, given that it is all 'just' physics and the inescapable fact that we cannot see some of the colors that hummingbirds see? The fact that light does what it does when it hits certain structures is an objective process. Iridescence can be found in nature in the absence, to us, of any behavioral 'usefulness'. For example, the iridescent inside of some abalone shells can be seen only when the shells are opened (and hence when they are dead) and must simply be a result of the shell's surface structure. Equally, many beetles have iridescent wing cases, especially those species that inhabit areas that are damper and/or with fine soil particles. In this case iridescence is not thought to be a display, but most logically a coincidental effect of the angled grooving on the wing cases that has evolved to minimize the problems of dust and water sticking to them—a bit like the grooves in a wet-weather tire.

ABOVE **White-tailed Starfrontlet** *Coeligena phalerata* (*left* | Colombia); **Black Jacobin** *Florisuga fusca* (*right* | Brazil);
BELOW **White-necked Jacobin** *Florisuga mellivora* | Colombia

Although iridescence occurs frequently in the natural world, especially in insects such as orchid or euglossine bees (*Exaerete* sp.) (*left* | Costa Rica) and on the inside of shells such as abalone (*right* | Australia), its precise purpose is still poorly understood.

However, there are also some beetle species that appear to have taken advantage of this coincidental iridescence and evolved to use it for display purposes.

In the case of hummingbirds, the main purpose of their often exquisite coloration is for display, both sexually and territorially. Iridescent colors, by the laws of physics, are directional within a narrow range. The dramatic changes in color and/or intensity caused by a slight change in viewing angle are evidence of this, changes in intensity (or brilliance) being especially obvious. This limited viewing angle suggests that hummingbirds may consciously use iridescence as a directional signal, to single out a particular female and, at the same time, avoid alerting other males to their presence. Further evidence of the importance of directional signaling is perhaps provided by their specially angled barbules, which assist in directing light. For example, in display, a male Anna's Hummingbird raises its throat and crown feathers and moves its head to create a flash of iridescence. Furthermore, on sunny days, males will often orient their diving display flights precisely to create a perfect angle towards the sun so that the full effect of this iridescence is directed towards the prospective mate.

Whether you choose to ponder these incredible light-bending nanostructures from a physical, spiritual or aesthetic perspective, the reality remains the same. These miniscule structures are incredibly complex and need to be positioned very precisely to work—yet there is an almost flawless consistency within species which boggles the mind. Furthermore, the fact that the positional accuracy of melanosomes is carried in the species' DNA is, apart from being thought-provoking, perhaps an indication as to how important color and iridescence is in signaling high-end genetic fitness and ultimately in the selection of a mate.

FACING PAGE The physics of color and the effect of iridescence are truly astonishing. Just look at the colors in the gorget of this **Rainbow-bearded Thornbill** *Chalcostigma herrani* and consider the fact that since some ultraviolet colors may also be involved, the overall effect, as perceived by another hummingbird, is likely to be something that we cannot even begin to imagine. | Ecuador

Sexual Dimorphism

The sexes of most hummingbirds look different (and hence a species is sometimes referred to as sexually dimorphic—two morphs, or forms), often dramatically so, with the male much more colorful than the female and in some species adorned with dramatic feather shapes. Since female hummingbirds have sole responsibility for nest-building and rearing the chicks, they benefit from being more camouflaged in order to minimize the risk of predation when incubating eggs or visiting the nest. Males, on the other hand, need to impress the female in order to encourage her to mate, and in some species the more brightly colored or extravagantly adorned he is the greater is his chance of being successful. Notwithstanding the plumage differences, the structure of the bill of the two sexes is usually similar, reflecting the relationship that the species has with favored flowers, although females tend to have a slightly longer or stronger bill than males. In some hummingbirds, however, such as the hermits and other members of the subfamily Phaethornithinae, and the caribs, the sexes look similar. This suggests that for these species there is no advantage to be gained from being brightly colored or particularly well camouflaged.

ABOVE The two sexes of the **Green-throated Carib** *Eulampis holosericeus* are very difficult to tell apart but females tend to have a slightly longer, more decurved bill | Martinique; BELOW **Green-crowned Brilliant** *Heliodoxa jacula* ♂ (*left*); ♀ (*right*) | Costa Rica

TOP **Black-crested Coquette** *Lophornis helenae*—♂ (*left*); ♀ (*right*) | Colombia
BOTTOM **Grey-tailed Mountain-gem** *Lampornis cinereicauda*—♂ (*left*); ♀ (*right*) | Costa Rica

When fledgling hummingbirds first leave the nest, males and females look very similar, their plumage generally resembling that of their mother. In most species, however, the fledglings soon start to molt into adult plumage and males gradually develop their often distinctive colors, a process that can take several months. This is most easily appreciated in species such as the **Snowcap** *Microchera albocoronata* that are clearly sexually dimorphic—immature ♂ (*above left*); adult ♂ (*above right*); ♀ (*below*). | Costa Rica

FACING PAGE The males of most species of hummingbird are strikingly colored and/or adorned to attract a mate. In some species, however, such as the **Purple-throated Carib** *Eulampis jugularis*, male and female look very similar, perhaps indicating that there is no particular advantage to the female in being camouflaged when breeding. | Martinique

Breeding: Continuing the Line

The day-to-day concerns for a hummingbird, as with any other bird, are staying healthy and avoiding predation. In the long term, however, the goal must be to reproduce and continue the line. Whether the need to self-perpetuate is driven by individual genes (the selfish gene theory), or by the birds themselves making a conscious decision to produce offspring, might be debatable. But either way, a hummingbirds' life-cycle ultimately revolves around finding a mate, nesting, hatching eggs, and rearing chicks to fledging and beyond.

Male hummingbirds, whatever method they may use to impress females, are polygynous: they mate with several females and remain with them for long enough only to fertilize a clutch of eggs. The male does nothing at all to help rear a family. The female does all the work, and the survival of the young is completely dependent upon her. In the case of a very few species, such as Green Hermit and Rufous-breasted Hermit, males may be seen close to a nest and appear to defend it against possible predators. Males of other species have occasionally been observed to attack a predator close to a nest, but this might have been coincidental rather than a deliberate defense of the nest.

In hummingbirds, as with most birds, males are more brightly colored than females, more likely to exhibit their colors in courtship displays and more likely to sing, both to attract a mate and to warn off rivals. Females cannot afford to show beautiful patterns or decorative plumes that so easily attract attention. Instead, they have to sit still on a nest for long periods, incubating their eggs, and later return repeatedly to feed a hungry family. They must, if they can, go unnoticed as they have no other viable defense against predators. Female hummingbirds may sometimes show a 'shadow' of the males' bright patterns and colors, although these are probably of no value and may even be a hindrance.

FACING PAGE In many hummingbirds, such as the **Violet-tailed Sylphs** *Aglaiocercus coelestis*, the male is much more conspicuous than the female (*inset*), reflecting the different roles of the pair when breeding. | Ecuador

BELOW The nests of most 'typical' hummingbirds, such as that of the **Purple-crowned Fairy** *Heliothryx barroti* ♀, may catch the human eye as an obvious irregularity or 'bump' on a shoot, twig or branch. Although they may appear improbably delicately balanced, such nests are invariably firmly anchored to ensure that they stay in place even when exposed to strong winds and heavy rain. | Colombia

This seems difficult to explain, but plumage patterns do take a considerable time to develop and the process may begin even before the sex of the embryo is determined inside the egg.

Females are often slightly longer-billed than males, probably to allow access to a wider range of food plants. Where there is a restricted range of nectar sources, males need to defend their favored flowers vigorously to ensure that they have access to a reliable food supply, chasing off any competitor, including females of their own species. The females may therefore have to search more widely and feed from different plants in order to obtain the nectar that they need. Females also require calcium to form eggshells, and a longer bill may allow them to feed on a slightly different range of foods that contain this essential mineral. In only a few hummingbirds, such as the caribs, do the male and female look nearly identical, although even in this case females have a slightly longer bill than that of males.

A male hummingbird is, it seems, more likely to attract a mate if he is more brightly colored than other males, the colors, coupled with the intensity of his courtship displays, presumably acting as a statement of 'quality', of fitness to father a brood, or possibly just his 'beauty' (see *page 60*). The number of males that actually fall short and remain unmated is, however, hard for us to assess. While the subtle differences between individual males may be obvious to a creature as minute as a female hummingbird, or a rival male, they are hard to judge from a human standpoint. It is rather like trying to imagine an alien, several hundred times bigger than us, appreciating the minuscule variations of facial expressions to which we humans are so sensitive.

Attracting a mate

Songs and leks

In most species of hummingbird the males sing individually, either from a particular songpost or display ground or as they go about their normal daily activities. A typical song, which is usually a subtle and discreet performance, is a simple short note followed by a longer one, repeated over and over for many minutes. It generally functions both as a warning to competitors to keep away from a feeding territory and as a means of attracting females. Male hummingbirds may make subtle adjustments to their song depending on the situation, but in general it seems that one type of vocalization serves both purposes. An intruding male might fight or flee, but if a visiting female is ready to mate she will not fly off immediately. Instead, she will sit still until the male notices that a potential mate has arrived and begins to respond accordingly—the pair quickly mating and each bird then going its own separate way.

The males of some hummingbirds are not territorial and instead sing cooperatively to attract females, a strategy known as lekking (which is used also by other groups of birds, such as manakins). A group of males, normally just a few but occasionally as many as 20 or more, will gather at a communal, often traditional, display 'arena' (the lek) and create a continuous cacophony of sound, allowing individual birds to take time out to feed or preen while their associates continue singing. The noise from a lek, which is frequently a remarkably loud and repetitive series of single "*chip*" notes, may persist all day and in some instances for much of the year. The males of certain

To the human ear, the notes of a hummingbird's song are usually hard to hear unless at close range, when a scattering of less musical, throaty sounds may also be heard. But in certain species, such as the **Bumblebee Hummingbird** *Atthis heloisa* (*above left*), the song is very distinctive and far-carrying. This hummingbird, which is found throughout the mountainous areas of Mexico at an altitude of 1,500–3,000 m (4,900–9,800 ft), has a distinctive prolonged, high-pitched buzzing song that is given with its bill wide open from an exposed songpost. | Mexico.
Many hummingbirds, such as the **Snowy-bellied Hummingbird** *Amazilia edward* (*above right*), vocalize in defense. | Panama

Hermits are much less brightly colored than many other hummingbirds, their plumage tending to be predominantly brownish or greenish, some species also having darker brown, orange or black markings. The color of the bill, especially of the lower mandible and the inside of the mouth, may be yellow, orange or reddish, and can be eye-catching when it is emphasized in the male's open-billed, calling display. ABOVE **Great-billed Hermit** *Phaethornis malaris* (*left*) | Ecuador; **Streak-throated Hermit** *Phaethornis rupurumii* (*right*) | Brazil; BELOW **Reddish Hermit** *Phaethornis ruber* | Brazil

Hummingbirds that display by lekking

This list provides an indication of the range of hummingbird species that display by lekking. It is based on published figures that summarize the number of males involved, but is not comprehensive, as other species are also known to form leks. The fact that figures are not available for almost half of these species indicates just how much there is still to be learned about this fascinating behavior.

Species	Males	Species	Males
Crimson Topaz	2–20	Scaly-breasted Sabrewing	up to 8
White-tipped Sicklebill	3–4	Wedge-tailed Sabrewing	up to 15
Band-tailed Barbthroat	3–5	Grey-breasted Sabrewing	2–4
Pale-tailed Barbthroat	?	Rufous Sabrewing	2
Little Hermit	?	Violet Sabrewing	?
Minute Hermit	?	White-tailed Sabrewing	2–4
Grey-chinned Hermit	?	Swallow-tailed Hummingbird	?
Reddish Hermit	?	Black-bellied Hummingbird	up to 5
Sooty-capped Hermit	3–6	White-tailed Emerald	2–5
White-bearded Hermit	?	Coppery-headed Emerald	3–6
White-whiskered Hermit	?	Snowcap	3–6
Green Hermit	12–16	Rufous-tailed Hummingbird	2–4
Tawny-bellied Hermit	?	Cinnamon Hummingbird	?
Straight-billed Hermit	?	White-bellied Emerald	3–6
Long-tailed Hermit	'exploded' 20–25	Blue-chested Hummingbird	usually 5
Great-billed Hermit	?	Charming Hummingbird	?
Brown Violet-ear	?	Golden-tailed Sapphire	5–10
Green Violet-ear	3–5	Blue-throated Goldentail	up to 10
Tooth-billed Hummingbird	?	White-eared Hummingbird	?
Green-breasted Mango	?	Amethyst-throated Hummingbird	?
Marvelous Spatuletail	up to 3	Anna's Hummingbird	'exploded'?
Green-crowned Brilliant	?	Broad-tailed Hummingbird	up to 3
Violet-headed Hummingbird	up to 10	Calliope Hummingbird	'exploded'?
Green- / Violet-crowned Plovercrest	2–7	Rufous Hummingbird	?

The purpose of a lek is to attract watching females—a strategy adopted by the males of many hummingbird species, such as these **Great-billed Hermits** *Phaethornis malaris*, which gather together to display, boosting their individual efforts by a cooperative endeavor that multiplies the effect of individual songs and displays. Leks may be large or small, concentrated or dispersed, depending on the species.

species, particularly some of the hermits, tend to stay within close visual contact, but those of others, such as the violet-ears, are more dispersed although still within audible range of one another; the latter situation is usually referred to as an 'exploded lek'. Once a female has been enticed to visit the lek and has appraised her prospective partners, mating takes place either on the male's songpost or close by. Quite how she chooses her partner is not entirely clear, but it may be based on subtle differences in the quality and duration of an individual male's song. Over the years, younger males perhaps learn how to improve their repertoire by listening to older birds.

Adornments and Display

The type of display used by courting male hummingbirds varies among species and often involves particular actions and postures that enable the birds to show off their finery to best effect. Crests are fanned and iridescent gorgets 'flashed' to focus one-directional beams of dazzling color at approaching females, or towards rival males. The hermits, which lack iridescence, fan their tail and open their bill to show bright colors inside the mouth. Indeed, if sunlight shines through the thin, colored skin at the gape, the result can be spectacular—like a sudden flash of color through a stained-glass window, or a tiny lamp being lit in the forest gloom.

Exotic adornment, rather than color, is most strikingly apparent with the male Marvelous Spatuletail. His tail feathers are reduced in number to just four (rather than the ten of all other hummingbirds), two of which have a long, wire-like shaft tipped with a broad, rounded, black 'racket' or spatula. These 'wires' are crossed at the base and curl forward in front of the displaying bird, in a wide 'U' shape. He waves his rackets as a female approaches and excitement builds—then leaves his perch and moves to-and-fro, while hovering like a tiny drone on rapidly fanned wings. The upright body reveals a broad, black central streak, and the middle tail feathers are angled down in a fine point: the rackets are fully spread and upswept, level with his head. The bird now looks like someone standing tautly upright, on tip-toe, arms out wide at head height while holding a tennis racket in each hand, and turning constantly to face his intended. After a few seconds, understandably, he is exhausted! (See illustration on *page 92*.)

When displaying, the males of some species of hummingbird add sounds made by their feathers to their limited vocal repertoire. In steep, zooming dives, tail feathers twist and vibrate, making a hum, or a buzz, or a short musical 'chirp' as specially shaped feather tips twist and buckle under the stress. On a few species, the tail feathers rattle against each other with a dry buzzing sound. Depending on the species, the tail feathers that

While striking patches of bright color are sufficient 'adornment' for many species when displaying, others have evolved equally dramatic feather structures. In some instances, males have both strong colors and eye-catching feathers, such as on this **Black-crested Coquette** *Lophornis helenae*. | Costa Rica

The striking long ear-tufts and raised crest of the male **Tufted Coquette** *Lophornis ornatus* are clearly intended to impress. | Trinidad

make these sounds may be the inner or outer ones, with either the broader inner web or the tip of the vane vibrating, each producing its own special sound by mechanical means. This is a technique that is used also by some other displaying birds, such as snipes and manakins, and in the insect world it is utilized by grasshoppers.

The displays that generate these sounds are truly remarkable, especially in the case of the smaller species of hummingbird. In steep dives and 'wing-tearing' recoveries, the bird may be subject to a force of up to 10 G—feeling ten times its own weight and having a huge pressure exerted on all its internal organs; to put this into perspective, fighter pilots have to be specially trained to withstand 9 G. A male Anna's Hummingbird, for example, starts at a height of up to 30 m (100 ft) and dives down at full speed, wings whirring to increase acceleration, and then, with great accuracy, swoops past a female or a key food source in a high-speed curve. Its dives have been timed at 95 kph (60 mph), and at the bottom of the dive the tail is briefly spread and produces a loud 'chirp' from its vibrating feathers. Measured as almost 385 body lengths per second, this speed makes Anna's Hummingbird the fastest vertebrate on earth relative to its size.

The dives made by some hummingbirds may be almost too fast to follow. A male Peruvian Sheartail, for example, has a long (7-cm/3-in), ribbon-like, black-edged, silvery-white tail that trails behind him as he dives, making a rattling, fluttering sound as the darker feather tips vibrate. Although the finer details are difficult to appreciate with the naked eye, even through binoculars, slow-motion video has been used to reveal the intricacies of its display: rather than trailing like a rippling ribbon, the tail feathers are instead spread straight out to the side, both during an initial series of sideways shuttles and during the long, fast, angled dive.

Males of other species have similar displays but use long, horizontal, semicircular flights, shooting past the perched female several times, or make shorter, more rapidly repeated dives, like a series of bouncy leaps. A male Allen's Hummingbird adds a sequence of wide, fast, pendulum-like side-to-side swings before the final dive; with the feathers in his wings and tail all vibrating at once he is able to make three different sounds at the same time. The tail feathers of a male Rufous Hummingbird create their own distinctive chattering sound, but each dive leaves the feathers dishevelled and, as he pulls up from his dive, he has to shake them back into place before flying off or starting over again.

Some male hummingbirds have a so-called 'shuttling display' that often precedes dives and steeper, dashing swoops, sometimes performed even while they are singing, and in which they whirr their wings and move jerkily from side to side. Male Calliope Hummingbirds perform a particularly rapid shuttling display, head-on to the female, with the colorful throat feathers fully

Male **Anna's Hummingbirds** *Calypte anna* take aerial displays to the extreme, combining rapid dives and swoops with 'instrumental' sounds created by the outspread tail feathers, which vibrate as the bird pulls out of its headlong plunge very close to a female.

The **Marvelous Spatuletail** *Loddigesia mirabilis* reveals marked sexual dimorphism in the structure of the tail: the male seeks to impress the female by swinging and shaking his long, wire-like feathers, each of which has a broad 'spatula' or racket at the tip.

extended in a broad fan, while creating a low, rhythmic "*brr-brr-brr-brr-brr*" buzz. Slow-motion video has revealed that the rapid wingbeats, which are merely a blur at normal speed, have minute pauses (or at least momentary reductions in speed) between each "*brr*" of the buzz. Similar footage of shuttle displays by other species, such as Oasis Hummingbird, shows similar brief pauses, in which the male actually closes his wings periodically during each flight.

Males of the evocatively named Black-tailed Trainbearer also make strange sounds when displaying. Their straight and slender tail, which is more than twice the length of the head and body, tends to be split by the breeze into a deep fork, or is gently upswept in one smooth curve as the bird hovers to feed. But when it is displaying, it 'claps' its tail in a short, sharp, accelerating rasping "*tp-tp-tr-rrrrp*" that sounds very much like a branch creaking in the wind.

Males of other hummingbirds, such as the Wire-crested Thorntail, display by hovering in front of a female, tail spread and crest erect, and sway back back and forth in an attempt to impress. Hummingbirds clearly have a remarkable range of display strategies, some involving specially adapted feathers or adornments and spectacular flights. Such displays may deter rivals, but ultimately their purpose must be to attract a mate.

A male **Wire-crested Thorntail** *Discosura popelairii* displays to a perched female. | Ecuador

Adornments and Display

Male hummingbirds exhibit a range of adornments that are used in display, ranging from crests and ear-tufts, to 'puff-legs' and colorful, and in some species extravagantly shaped, tails. Some examples of hummingbirds that have such adornments are shown *here* and on *pages 96–97*.

FACING PAGE **Buff-tailed Coronet** *Boissonneaua flavescens* | Colombia; ABOVE **Antillean Crested Hummingbird** *Orthorhyncus cristatus* (*left* | Martinique); **Brown Violet-ear** *Colibri delphinae* (*right* | Ecuador; BELOW **Ruby-topaz Hummingbird** *Chrysolampis mosquitus* | Brazil

ABOVE **Hooded Visorbearer** *Augastes lumachella* | Brazil; BELOW **Racket-tailed Coquette** *Discosura longicaudus* | Brazil; FACING PAGE **Rufous-crested Coquette** *Lophornis delattrei* | Peru

Aggressive Behavior

Plants do not constantly secrete nectar and once a hummingbird finds a productive patch of its favored flowers it will often defend it to ensure a regular food source. In some areas as many as 15 species compete for nectar, but three or four species are more likely. In certain species there are dominant males, in others the birds accept a subordinate role and feed accordingly, while in some species individuals hide nearby and wait for a chance to feed when the dominant bird is elsewhere. These strategies, despite aggression and territorial defense, allow several species to live in one area. In the breeding season, males of species such as the Ruby-throated Hummingbird will initially drive off both males and females, but once a female has been accepted as a mate she will later be allowed free access to the male's feeding territory. This is particularly beneficial when the female is incubating eggs or brooding small chicks and cannot afford more than brief feeding forays.

BELOW **Green-crowned Brilliants** *Heliodoxa jacula* ♀s| Costa Rica; FACING PAGE **Chestnut-breasted Coronets** *Boissonneaua matthewsii* (*top* | Ecuador); **White-tailed Starfrontlet** *Coeligena phalerata* (*bottom, upper bird*) and **Sparkling Violet-ear** *Colibri coruscans* (*bottom, lower bird*) | Colombia

Nests and nest-building

Once it is time to construct her nest, a female hummingbird will typically select a suitable nest site near a good feeding area, repeatedly visiting and settling on a suitable branch, as if testing its qualities. Female hermits, however, do not nest close to feeding sites, but instead select hanging foliage, testing the strength of the leaves by clinging to them with their minuscule feet.

Hummingbird nests are of many shapes but all are tiny structures, seemingly incapable of holding even a minute brood of hummingbirds. The majority of species build cup-shaped, open nests, averaging barely 5 cm (2 in) across, and placed either on top of a branch and extending down on each side, like a tiny saddle, or on a large leaf or in the fork of a horizontal twig. A few of these nests seem able to survive even in apparently impossible locations, such as on overhead wires and even clotheslines in backyards. The nests of some other species swing more freely from drooping liana vines or from beneath a large leaf that offers shelter, and a few hummingbirds plaster their nests on to smooth rocks. Female sylphs are unusual in that they build domed nests, and indeed male sylphs also build semi-completed structures but these are used only as shelters at night. Although most hummingbirds nest singly, some, such as the Rufous-tailed Hummingbird, may breed in loose colonies, the females sometimes even stealing nest-building material from other nests nearby.

A hummingbird nest needs to be easily accessible, as the female must be able to fly directly to it rather than approaching by creeping through foliage as many other birds do. Nevertheless, most are built in a more or less sheltered location and often close to water, which mitigates local temperature fluctuations and increases humidity. Nests even of one and the same species, may, however, be positioned almost anywhere from near ground level to high up in trees.

Female hermits usually locate their nest on the inside of a palm leaf, but the building process is especially difficult as initially she has to work while hovering, since there is no perch available. The foundations of the nest must be stuck to the leaf with a mixture of saliva and

Amethyst Woodstar *Calliphlox amethystina* ♀ | Brazil

nectar, and only when she has created a substantial mound can she begin to grip with her feet and complete the construction. A distinctive characteristic that the nests of all hermits share is a long, pendulous 'tail' of roots, leaves and downy plant material, occasionally weighted down with bits of earth. The 'tail' on the nest of the Sooty-capped Hermit is a remarkable example, being suspended on a single woven strand of spiders' web attached to the cup-shaped nest at one point along the edge: the long 'tail' acts as a balancing counterweight, keeping the nest upright, even in a wind. The long, loose tail on a hermit's nest can be surprisingly eye-catching to a person wandering by, but may be overlooked as simply a random piece of hanging vegetation by a passing predator. In particularly wet regions, hermits' nests are loosely made so that water from frequent showers can pass right through very quickly without cooling the eggs.

The nests of 'typical' hummingbirds have a tough outer layer and a softer lining, and may be overlain with pieces of moss, lichen, leaves and roots that are bound together with spiders' web, creating excellent camouflage; the rim is slightly turned in to help prevent the chicks from falling over the side. A few species may add what appear to us to be bits of decoration, such as scraps of red flowers, but the purpose of this is not obvious. High-altitude hummingbirds tend to build larger and bulkier nests than those of their lowland relatives, often in dense thickets of low vegetation or in various natural or man-made cavities and hollows in banks and roadside cuttings, even beneath bridges or in culverts, such sites offering greater security and shelter for the eggs and chicks. At particularly high elevations, nighttime can bring glazed ice or drifts of fresh snow and some form of shelter is therefore essential.

Seemingly fragile nests can be surprisingly tough. In the case of some species, such as the Rufous Hummingbird, nests from the previous year may be reused, and not always by the same individual. Sometimes, such as in the case of Anna's Hummingbird, an old nest, rather than being reused, will instead be picked apart for material to be incorporated in a new one. Other old nests may simply be used as a foundation for a new one.

LEFT **Rufous-tailed Hummingbird** *Amazilia tzacatl* ♀ | Mexico; RIGHT **Rainbow-bearded Thornbill** *Chalcostigma herrani* ♀ | Colombia

Nests

Hummingbirds' nests may be small but are remarkable constructions and are often built in unlikely places. The hermits (*shown here*) build a tapered nest that extends the drip-point of slender leaves, looking to the casual eye like a random accumulation of plant debris. The long 'tail' of the nest acts as a balancing counterweight, and to improve stability even further a relatively heavy piece of material, such as a lump of earth, is occasionally attached to the tip.

Great-billed Hermit *Phaethornis malaris* ♀ | Colombia

FACING PAGE **Reddish Hermit** *Phaethornis ruber* ♀ (*left* | Guyana); **Hook-billed Hermit** *Glaucis dohrnii* ♀ (*right* | Brazil)

The nests of 'typical' hummingbirds are quite different from those of the hermits. A female Amethyst Woodstar, for example, builds a tiny cup, secured by a wrap of fine fiber to the top of a branch and decorated with lichens for camouflage, while a Speckled Hummingbird relies instead on dense cover to shelter her nest and hide it from predators. Although the nests of the hummingbirds shown *opposite* may appear to have been built in an exposed, vulnerable location, they are wonderfully camouflaged with plant materials.

ABOVE **Amethyst Woodstar** *Calliphlox amethystina* ♀ | Brazil; BELOW **Speckled Hummingbird** *Adelomyia melanogenys* ♀ | Colombia

ABOVE LEFT **Shining Sunbeam** *Aglaeactis cupripennis* ♀ | Ecuador; ABOVE RIGHT **Tourmaline Sunangel** *Heliangelus exortis* ♀ | Colombia; BELOW **Red-billed Emerald** *Chlorostilbon gibsoni* ♀ | Colombia

Eggs and chicks

With one exception, hummingbirds lay a clutch of two oval, unmarked and slightly shiny white eggs, the 'oddity' being the Giant Hummingbird, which lays just one egg. Nests with three eggs do occur from time to time, but this appears to be the result of two females using the same nest. Although this phenomenon is most frequently encountered in the hermits, particularly those in the genera *Glaucis* and *Threnetes* (barbthroats), it is occasionally found also in 'typical' hummingbirds of the genera *Colibri* (violet-ears), *Amazilia* (which contains a wide range of species) and *Calypte* (Anna's and Costa's Hummingbirds). The reason for such 'egg-dumping' is difficult to explain, but the behavior is not infrequent within a wide range of other bird species, particularly wildfowl.

Hummingbirds lay their eggs at an interval of two days, the first sometimes even before the nest is properly complete. Incubation usually begins after the second egg is laid, in order to ensure synchronized hatching, but in some cases the female starts to sit early and her eggs hatch a day or two apart. Although the eggs are tiny, those of the Vervain Hummingbird of Hispaniola and Jamaica being the smallest of all birds' eggs at less than 10 mm (0.39 in) in length and weighing only around 0.37 g (0.013 oz), in the case of smaller hummingbird species each egg may still be more than 15% of the female's body weight. Surprisingly, the incubation period of around 14–19 days is longer than that for most other small birds, and for hummingbirds living in colder climes, either at high latitudes or at high altitudes, it may be up to 23 days. (For comparison, the incubation period of a typical small warbler, such as the Chestnut-sided Warbler, is 11–12 days.)

When it comes to incubation, a tiny female hummingbird, which would normally feed every few minutes in order to meet its high energy demands, could have a very real problem: heat has to be transferred to the eggs to allow development of the embryos, and she must incubate them, with no assistance from a male, for up to 90% of daylight hours, as well as through the night. She is, however, able to reduce her body temperature by some 9°C (16°F), thereby saving about 50% of her energy and therefore not needing to feed so frequently. Even so, the fact that a hummingbird is able to sit for so long on a nest before the eggs hatch is quite astonishing. From time to time the eggs must be turned, which a hummingbird does—again uniquely—by hovering over the eggs and using its feet.

The chicks hatch virtually naked except for two short tracts of down on the back, and they are blind for the first five or six days. The female feeds them by regurgitating a mixture of nectar and tiny fragments of insects and other small invertebrates, stimulating them to open their bill by touching their closed, bulging eyes. She approaches her nest on a variable spiraling or

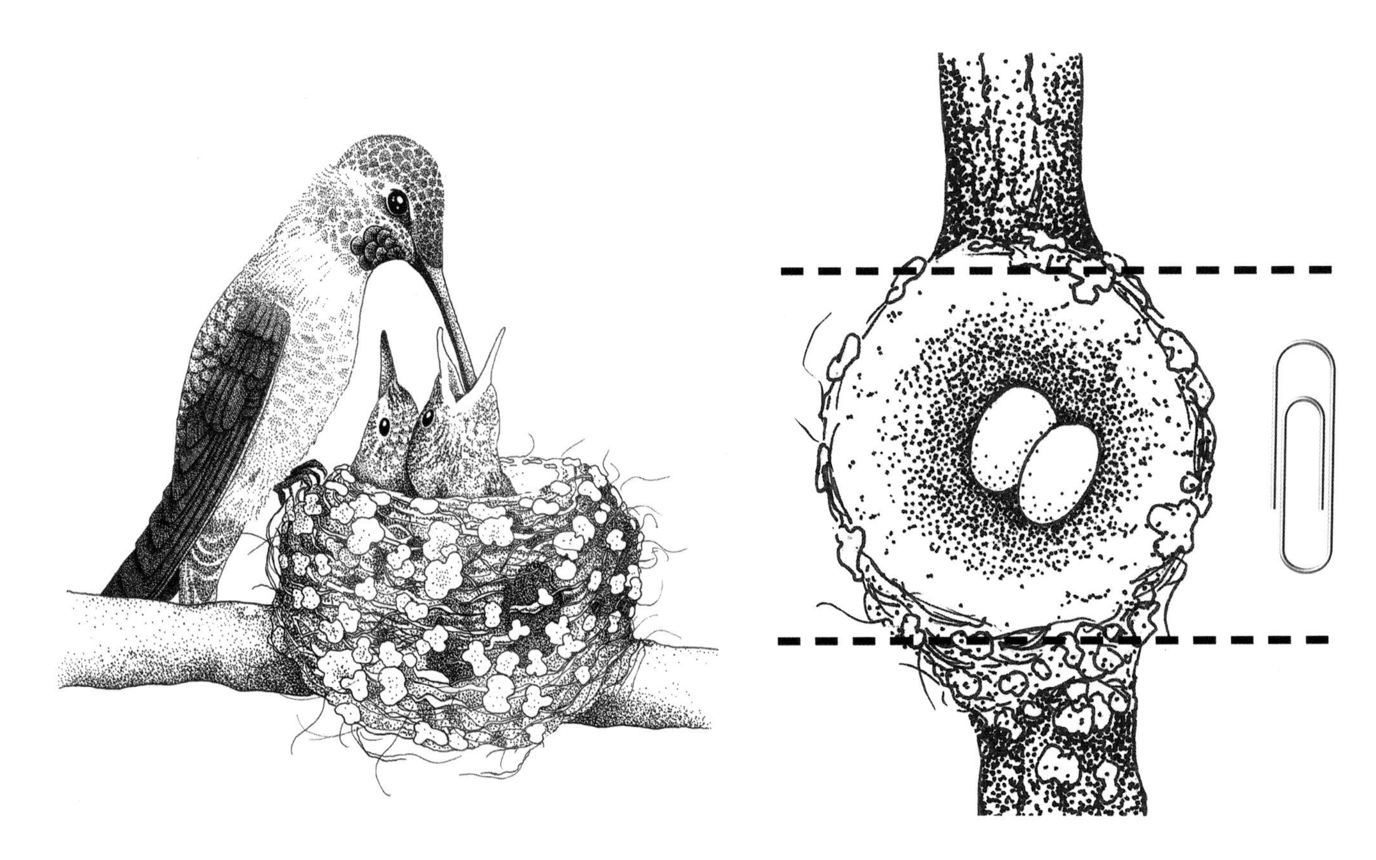

Anna's Hummingbird *Calypte anna* nest and eggs shown at approximately life-size.

Hummingbird Predators

Hummingbirds have lightning-fast reactions but predators can be extremely difficult to see even at close range (such as the viper below). More vulnerable, obviously, is the nest with its eggs, chicks and incubating or brooding female, literally a 'sitting target' for a variety of predatory birds, reptiles and mammals.

Hummingbirds' main predators include snakes such as the **Eyelash Viper** *Bothriechis schlegelii* (*above*), and birds such as **Keel-billed Toucan** *Ramphastos sulfuratus* (*below*). | Costa Rica

Andean Hillstar *Oreotrochilus estella* | Ecuador

White-eared Hummingbird *Basilinna leucotis* | Guatemala

zigzag path and often departs with a curling descent like a falling leaf, both of which actions seem likely to be predator-avoidance strategies.

At six to nine days of age, the minuscule chicks start to grow feathers and begin to sit on the edge of the nest. Once their eyes are open, the chicks beg for food only in response to the column of air being swept across them by the female's beating wings, the original down, still attached to the growing feathers, appearing to be particularly sensitive to such air movement. Unlike most young birds wanting to be fed, hummingbird chicks make no sound, the only exceptions being those species, such as the sylphs, that make dome-shaped nests. The female broods the chicks regularly until they are 7–12 days old, helping to regulate their temperature by brooding or shading, and she feeds them about twice an hour. The chicks grow fast, and within 10–12 days they are close to 80% of their final adult size.

Hummingbird chicks remain in the nest for between 26 and 40 days (longest in those species that live at high altitude), a surprisingly lengthy period for such small birds (Chestnut-sided Warbler chicks, for example, fledge after 10–11 days). The fledging success of hummingbirds is moderate, generally between 20% and 40%, but up to 60% for some high-altitude species and an exceptional 68% for the Andean Hillstar. In studies undertaken in tropical regions, most chick losses were due to adverse weather, with about a third lost to predation by birds, such as jays and toucans, as well as by reptiles and smaller mammals. In the high Andes, however, there are virtually no predators, and mortality due to predation is therefore reduced almost to nil. In North America the situation is reversed, with predation much the most important factor in fledgling mortality. Some hummingbirds, though, such as the Black-chinned Hummingbird, have come up with a cunning means of defense: often deliberately building their nest close to the nest of a bird of prey, sometimes even in the same tree. This seemingly paradoxical behavior is in fact beneficial to the hummingbird, as the presence of the raptors deters potential nest predators.

Once the chicks leave the nest they are fed by the adult female for a further 18–25 days, occasionally longer in the case of some tropical species. They initially stay in one spot, the female returning to feed them, but after a few days she calls loudly to the youngsters to attract them towards food, and gradually they learn which flowers to visit and which areas are especially rich in nectar. Despite this demanding schedule, the female may begin to build a second nest while still feeding young from the first, some even laying a second clutch almost immediately after the chicks from the first have fledged. Although most hummingbirds have one or two broods each year, a few species have occasionally been known to raise three.

Remarkably, for some species of hummingbird it has been found that the female may still be feeding large

FACING PAGE **Oasis Hummingbird** *Rhodopis vesper* ♀ feeding chicks | Chile

ABOVE These **Black-chinned Hummingbird** *Archilochus alexandri* (*left* | USA) and **Horned Sungem** *Heliactin bilophus* (*right* | Brazil) chicks (the latter being fed by the female) are just a few days old and will remain in their nest for about four weeks, a relatively long period for such a small bird.

BELOW It is extraordinary to watch a female hummingbird, such as this **Ruby-topaz Hummingbird** *Chrysolampis mosquitus*, feeding minute chicks, seemingly probing so deeply inside the chick's throat with her long, pointed bill that damage seems almost inevitable! | Brazil

ABOVE **Rufous-tailed Hummingbird** *Amazilia tzacatl* chick (*left* | Colombia); **Green-tailed Goldenthroat** *Polytmus theresiae* chicks (*right* | Brazil);
BELOW **Sparkling Violet-ear** *Colibri coruscans* ♀ feeding chicks | Ecuador

chicks in one nest while incubating eggs in a second. This was first noticed for the White-eared Hummingbird of Mexico in the 1930s, and it has since been studied in Rufous and Calliope Hummingbirds. It is now known that other hummingbird species that occur in North America also use this breeding strategy, and it has been speculated that the shorter breeding season in more northerly regions may encourage such overlapping nesting attempts. The incidence of 'overlapping broods' has not yet been recorded for the hummingbirds of tropical regions farther south, probably because conditions there are suitable for breeding year-round.

Most hummingbirds appear to be mature and breed when one year old, and the average lifespan of most species is probably six to eight years at best, although individuals can sometimes live for up to 14 years in captivity. The longevity of hummingbirds in the wild is, however, more precisely known for a few North American species, the Magnificent, Black-chinned and Buff-bellied Hummingbirds having exceeded 11 years, and the oldest, at 12 years two months, being a Broad-tailed Hummingbird. As with so much of hummingbird biology, however, the age to which most species can live remains unknown.

Although there are still many gaps in our knowledge—for a few species even the nest has not yet been described—the facts and figures pertaining to hummingbird breeding biology seem to be fairly consistent. These gaps will no doubt be filled as research continues, and it seems likely that there will still be some surprises in store.

Hybridization

Despite the suggestion that female hummingbirds actively select a mate, displaying males may pursue females aggressively and forced copulations are not infrequent. Occasionally, a male will even mate with a female of the wrong species, sometimes even from a different genus, and produce a hybrid. Although hybrid hummingbirds are not uncommon, they are infertile and do not persist beyond the first generation. Where subspecies, rather than full species, interbreed, the 'hybrids' are generally fertile and show characters that are intermediate between those of their parents. If the young are infertile, however, this suggests that the subspecies concerned are already at or close to full species level.

It seems that hybrid hummingbirds are most likely to occur when one or both parent species are rare and the population level low, reflecting difficulties in finding a mate of the same species. Nevertheless, some of the hybrids described have been between common species such as Costa's and Calliope Hummingbirds in North America. Hybrid hummingbirds usually have features that approach one or other parent more closely, but individuals that clearly have mixed features are extremely difficult to identify with any certainty. Indeed, female hybrid hummingbirds are especially likely to go unnoticed as they tend to lack distinctive characteristics. In rare cases a hybrid may not resemble either parent and may look like a different species altogether.

In the past, hybrid hummingbirds have often been described as new species, some of which are now known to account for species that were previously regarded as 'lost' (or extinct), including those described from skins in collections or, in some cases, from specimens used as decoration on robes and headdresses from ancient cultures. The Bogotá Sunangel is the most recent example of a 'species' that later proved to be a hybrid, and is one that has received considerable attention. This striking hummingbird is known only from a single specimen that was purchased in 1909 in Bogotá, Colombia, but there was no information as to where and when it was collected. In 1993 it was described as a new species, *Heliangelus zusii*, but some scientists remained doubtful until its status was 'confirmed' by DNA analysis in 2009. A later DNA study, published in 2018, revealed, however, that this bird was in fact a hybrid after all—between a female Long-tailed Sylph (widely distributed through the Andes) and a male of another, uncertain, species. A further twist in the saga came in 2011, before the second DNA study was undertaken, when an unusual hummingbird was discovered at the Rogitama Biodiversity Reserve in Boyacá, Colombia. This attracted much media attention, as there was speculation that it might represent the rediscovery of the Bogotá Sunangel. But this was not to be, as the bird in question was subsequently shown likewise to be a hybrid, with Long-tailed Sylph and Tyrian Metaltail the likely parents.

Compared with other bird families, the frequency with which hummingbird hybrids have been described as species in published lists is remarkably high, at 34 (information on each of these is included in Appendix 2, at the end of the chapter *Taxonomy: The BirdLife List of Species* on *page 275*). There are also another 30 'species' of hummingbird described in the published literature that are now known to be the same as another named species (a so-called synonym) or have been shown to be, or are suspected to be, aberrant or immature forms of another species (these are listed in Appendix 1 on *page 274*).

– o O o –

Having considered the various aspects of hummingbird biology—what they are, how they live and how they breed—we now turn to *where* they live.

The main image (*above*) is the so-called Monserrate hummingbird, that was discovered in 2020 at Cerro de Monserrate in Bogotá, Colombia. It is believed to be a hybrid between **Golden-bellied Starfrontlet** *Coeligena bonapartei* (*top left*) and **Blue-throated Starfrontlet** *Coeligena helianthea* (*top right*). This bird looks similar to Golden-bellied Starfrontlet but is darker overall, with a more richly copper-colored breast and a violet throat spot, perhaps indicating the presence of the genes of Blue-throated Starfrontlet. | Colombia

The main image (*above* | Colombia) shows the hummingbird that was the subject of considerable media attention when it was seen at the Rogitama Biodiversity Reserve, Boyacá, Colombia, in 2011. At the time there was speculation that it might represent the rediscovery of the Bogotá Sunangel but subsequent studies have shown it to be a hybrid, probably between a **Long-tailed Sylph** *Aglaiocercus kingii* (*top left* | Ecuador) and a **Tyrian Metaltail** *Metallura tyrianthina* (*top right* | Ecuador).

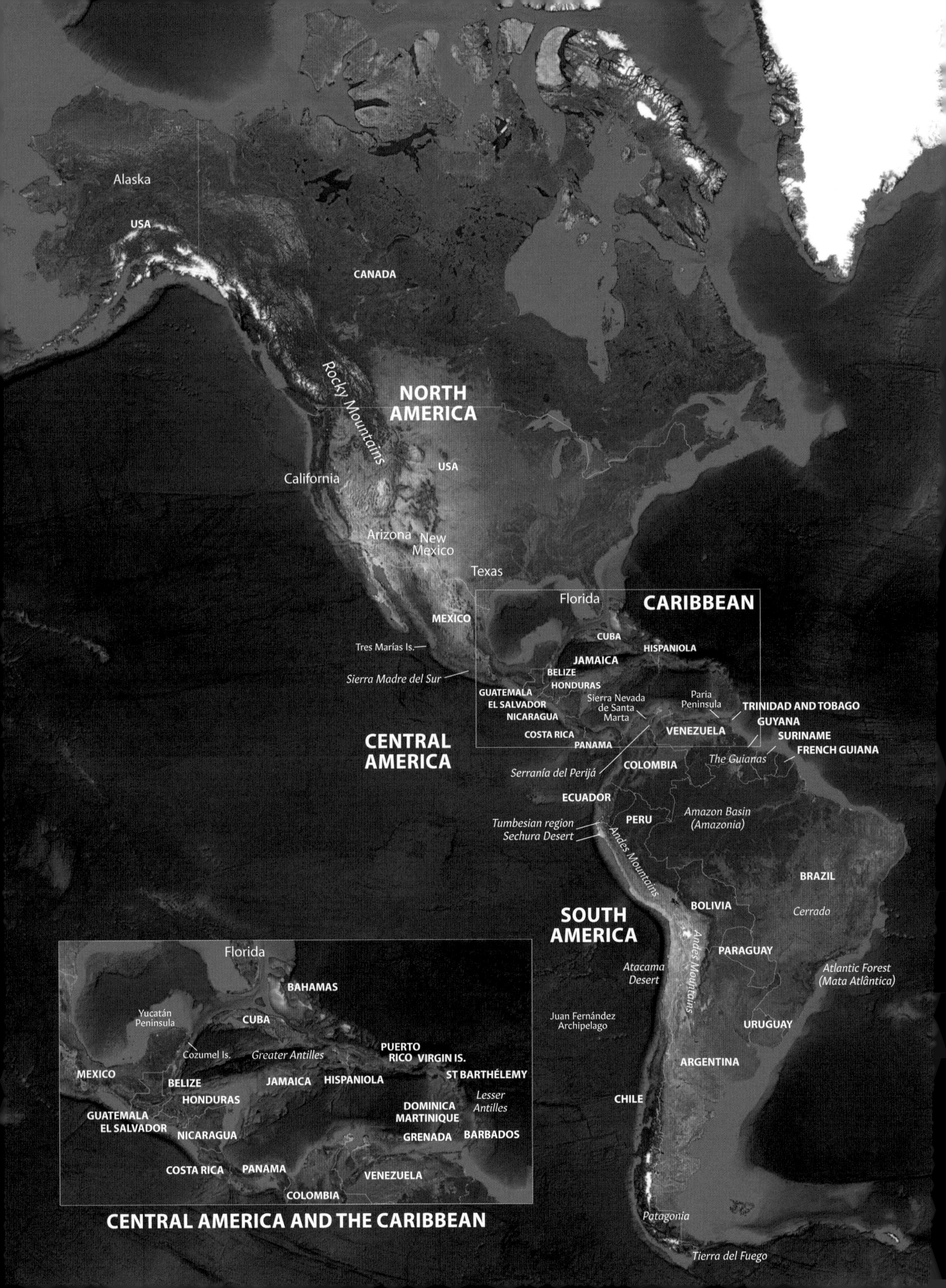

Alaska
USA
CANADA
Rocky Mountains
NORTH AMERICA
USA
California
Arizona
New Mexico
Texas
Florida
CARIBBEAN
MEXICO
CUBA
HISPANIOLA
Tres Marías Is.
JAMAICA
BELIZE
Sierra Madre del Sur
HONDURAS
GUATEMALA
EL SALVADOR
NICARAGUA
Sierra Nevada de Santa Marta
Paria Peninsula
TRINIDAD AND TOBAGO
GUYANA
COSTA RICA
VENEZUELA
SURINAME
PANAMA
FRENCH GUIANA
CENTRAL AMERICA
The Guianas
Serranía del Perijá
COLOMBIA
ECUADOR
PERU
Amazon Basin (Amazonia)
Tumbesian region
Sechura Desert
Andes Mountains
BRAZIL
BOLIVIA
Cerrado
SOUTH AMERICA
Andes Mountains
PARAGUAY
Atacama Desert
Atlantic Forest (Mata Atlântica)
Juan Fernández Archipelago
URUGUAY
ARGENTINA
CHILE
Patagonia
Tierra del Fuego
Florida
BAHAMAS
Yucatán Peninsula
CUBA
Cozumel Is.
Greater Antilles
PUERTO RICO
VIRGIN IS.
MEXICO
ST BARTHÉLEMY
BELIZE
JAMAICA
HISPANIOLA
Lesser Antilles
HONDURAS
DOMINICA
MARTINIQUE
GUATEMALA
EL SALVADOR
NICARAGUA
GRENADA
BARBADOS
COSTA RICA
PANAMA
VENEZUELA
COLOMBIA
CENTRAL AMERICA AND THE CARIBBEAN

Biogeography and Biodiversity:
The Hummingbirds' Realm

Hummingbirds are found only in the Americas, from Tierra del Fuego in the south to southern Alaska in the north. Many hummingbird species are able to thrive in a wide variety of habitats and have a widespread distribution, while others have evolved to take advantage of very specific conditions, with some of these species restricted to very small areas indeed. Furthermore, while some species are common, others are extremely rare, with a total known population in the hundreds. The primary factor behind hummingbird distribution is the distribution of the nectar-rich flowers they depend on to sustain their high energy requirements. To ensure this vital energy supply some species migrate (a few over long distances), while others move altitudinally, to take advantage of different flowering periods.

Moving broadly from north to south, this chapter presents an overview of this remarkably diverse family, giving an insight into the many different habitats in which hummingbirds occur and their often astonishing migrations.

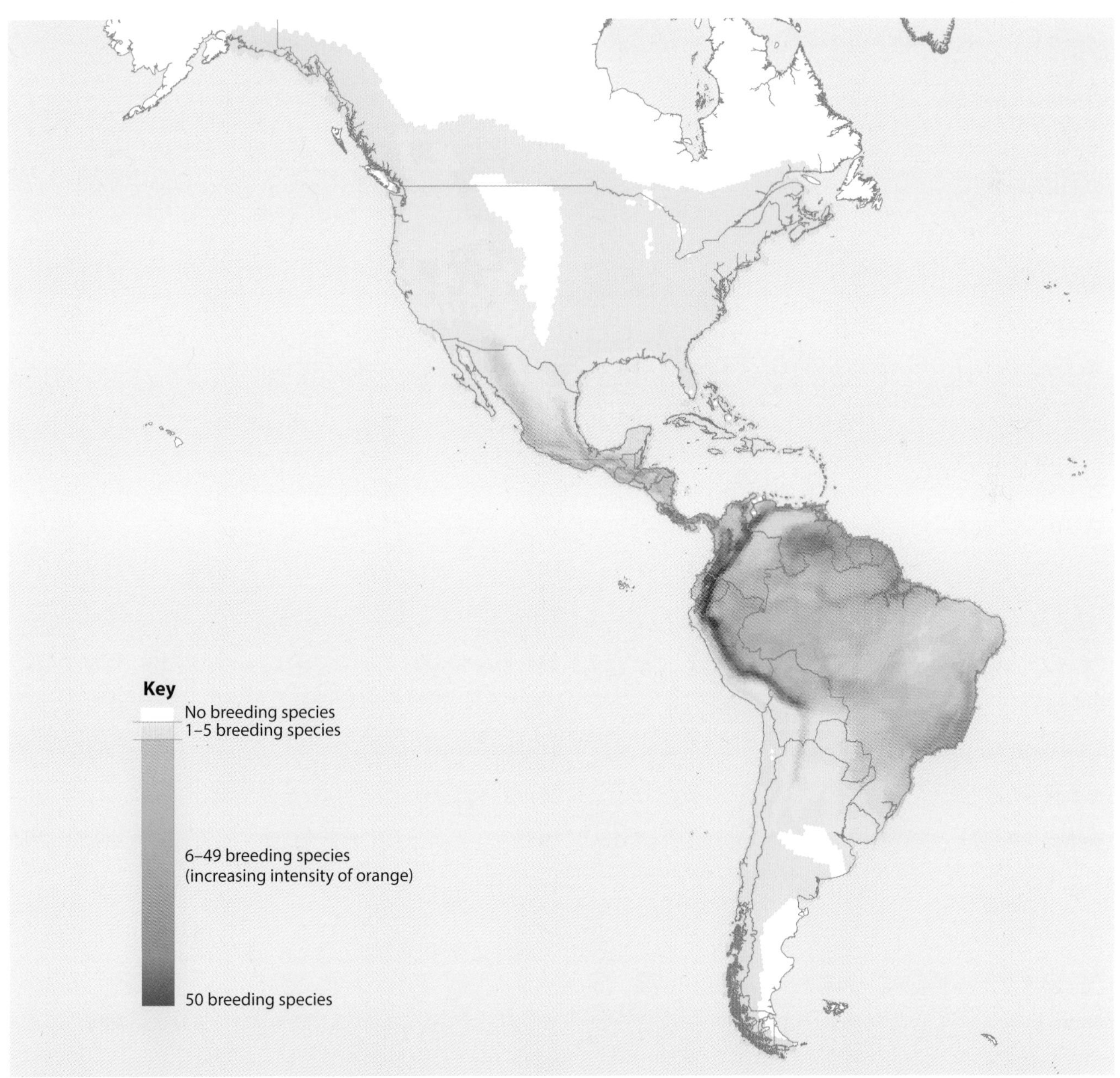

This 'heat map' illustrates the breeding distribution of hummingbird species across the Americas: the darker the shading, the greater the number of species. It highlights the impact of altitude, and in particular the Andes Mountains, on hummingbird diversity. The yellow areas indicate 1–5 species and the reddest up to 50 species, based on a hexagonal grid of 'cells', each of which equates to an area of approximately 2,615 km^2. [Source: BirdLife International and Handbook of the Birds of the World (2020) Bird species distribution maps of the world. Version 2020.1. Available at http://datazone.birdlife.org/species/requestdis]

The distribution and diversity of hummingbirds

This table summarizes the number of hummingbird species recorded in each country of the Americas and the number that breed only in that country (these species are termed breeding endemics). The number of vagrants (species that have been recorded as occasional 'strays' outside their usual or expected range) is also given, based on data available online via *Avibase* – The World Bird Database (see *page 277*). The figures follow BirdLife International taxonomy (see *page 241*).

NORTH AMERICA	**Regularly occurring species** (+ vagrants)	**Breeding endemics**	**% endemics**
Canada	**5** (+7 vagrants)	—	—
USA	**14** (+10 vagrants)	**1**	**7%**
THE CARIBBEAN			
All islands (including the Bahamas, but excluding Trinidad and Tobago, which are, geologically, part of South America)	**20** (+2 vagrants)	**17**	**85%**
CENTRAL AMERICA			
Mexico (including the islands of Cozumel and Tres Marías)	**60**	**13**	**22%**
Guatemala	**39**	—	—
Belize	**22**	—	—
El Salvador	**23**	—	—
Honduras	**41**	**1**	**2%**
Nicaragua	**36**	—	—
Costa Rica	**53**	**2**	**4%**
Panama	**58**	**2**	**3%**
SOUTH AMERICA			
Colombia	**161**	**18**	**11%**
Ecuador	**136**	**5**	**4%**
Peru	**132** (+1 vagrant)	**20**	**15%**
Bolivia	**79**	**2**	**3%**
Chile (including the Juan Fernández Islands)	**9** (+1 vagrant)	**2**	**22%**
Argentina	**28** (+5 vagrants)	—	—
Venezuela	**104**	**5**	**5%**
Guyana	**40**	—	—
Suriname	**34**	—	—
French Guiana	**33**	—	—
Brazil	**90**	**18**	**20%**
Paraguay	**20**	—	—
Uruguay	**6** (+2 vagrants)	—	—

The USA and Canada

The USA has 14 breeding species of hummingbird, five of which are found also in Canada. Across large swathes of southern Canada and the eastern USA, the Ruby-throated Hummingbird is the only member of the family that breeds there, while in the west and southwest Rufous, Black-chinned, Anna's and Calliope Hummingbirds are the most frequently encountered. The hummingbirds of this region occupy a wide range of habitats, from forests on the fringes of Arctic Alaska to the vast expanses of desert in the southwest. Many species are equally happy in urban situations and can be easily seen in parks, gardens and backyards, some regularly visiting feeders.

Although a few of the hummingbirds in the west of the region are at least partially resident—notably Anna's and Costa's Hummingbirds—most of the species that breed in the USA and Canada are either long-distance or short-distance migrants. The Ruby-throated Hummingbird, for example, undertakes a particularly arduous migration that involves a flight across the Gulf of Mexico. Equally impressive is the Rufous Hummingbird, which winters in Mexico and southwestern USA, and is renowned for having been recorded at the most northerly latitude of any hummingbird, vagrants having been found even on the western Alaskan islands. More detailed information on the migrations of North American hummingbirds is provided in *Epic migrations* (*page 174*).

Four of the breeding hummingbirds are essentially 'Mexican' species, the ranges of which extend north into the USA, where they are restricted mainly to the canyonlands of Arizona, New Mexico and western Texas.

The rarest breeding hummingbird in the USA is the Violet-crowned Hummingbird, which was first recorded breeding just north of the Mexican border in 1959.

In addition to the 14 regularly occurring species, ten vagrants have been recorded in the USA, three of which have made their way as far north as Canada. These vagrants (marked with an asterisk * if recorded in Canada) are: Green Violet-ear * (subspecies *thalassinus*, which is considered by some taxonomists to be a separate species, Mexican Violet-ear); Green-breasted Mango; Cinnamon Hummingbird; Berylline Hummingbird; Xantus's Hummingbird *; White-eared Hummingbird; Plain-capped Starthroat; Amethyst-throated Hummingbird *; Bahama Hummingbird and Bumblebee Hummingbird.

Sonora Desert | Arizona, USA

Widespread Hummingbirds of the USA and Canada

Ten of the 14 hummingbird species that occur regularly in the USA have a widespread distribution—shown *here*. The breeding range of five of these species extends into Canada (*), and three have occurred only as vagrants (‡). (Although Allen's and Buff-bellied Hummingbirds breed in the USA, they have not been recorded in Canada.) See the relevant entries in *The BirdLife List of Species* (*page 241*) for information on status and distribution.

RIGHT **Anna's Hummingbird** *Calypte anna* * | Canada;
CENTER LEFT **Black-chinned Hummingbird** *Archilochus alexandri* * | USA;
CENTER RIGHT **Broad-tailed Hummingbird** *Selasphorus platycercus* ‡ | USA;
BOTTOM **Ruby-throated Hummingbird** *Archilochus colubris* * | USA

TOP LEFT **Broad-billed Hummingbird** *Cynanthus latirostris* ‡ | USA; TOP RIGHT **Buff-bellied Hummingbird** *Amazilia yucatanensis* | USA; CENTER LEFT **Costa's Hummingbird** *Calypte costae* ‡ | USA; CENTER RIGHT **Calliope Hummingbird** *Selasphorus calliope* * | Canada; BELOW LEFT **Allen's Hummingbird** *Selasphorus sasin* | USA; BELOW RIGHT **Rufous Hummingbird** *Selasphorus rufus* * | USA

Localized Hummingbirds of the USA and Canada

Four of the 14 hummingbird species that occur regularly in the USA are essentially 'Mexican', the ranges of which extend just over the border; one has occurred as a vagrant in Canada (‡). See the relevant entries in *The BirdLife List of Species* (*page 241*) for information on status and distribution.

RIGHT **Magnificent Hummingbird** *Eugenes fulgens* ‡ | USA;
CENTER LEFT **Blue-throated Hummingbird** *Lampornis clemenciae* | USA;
CENTER RIGHT **Violet-crowned Hummingbird** *Amazilia violiceps* | USA;
BOTTOM **Lucifer Hummingbird** *Calothorax lucifer* is one of the few hummingbirds that has adapted to desert life. It inhabits arid canyons and foothills in mountainous areas, where it favors scrubby areas and is often seen feeding on flowering century plants (agaves). Lucifer Hummingbirds are in fact 'nectar thieves': they are so small that they are able to reach the nectar without actually pollinating the plant. | USA

Vagrant Hummingbirds of the USA and Canada

Ten species of hummingbird have occurred as vagrants in the USA, three of which have been recorded in Canada (‡). Four of these species are shown *here*, the others being Green Violet-ear (‡), Bahama Woodstar, Berylline Hummingbird, Plain-capped Starthroat, Cinnamon Hummingbird and Bumblebee Hummingbird. See *The BirdLife List of Species* (*page 241*) for information on status and distribution.

RIGHT **White-eared Hummingbird** *Basilinna leucotis* | Guatemala;
CENTER LEFT **Amethyst-throated Hummingbird** *Lampornis amethystinus* ‡ | Guatemala;
CENTER RIGHT **Xantus's Hummingbird** *Basilinna xantusii* ‡ | Baja Peninsula, Mexico;
BOTTOM **Green-breasted Mango** *Anthracothorax prevostii* | Costa Rica

The Caribbean—A Unique Suite of Hummingbirds

Between the southern tip of Florida and the Yucatán Peninsula of Mexico and the coasts of Central and South America, are the tropical islands of the Caribbean Sea. Their geological history is complex, but their relative isolation from the continental landmass, and in some cases from each other, has resulted in a wide range of plants and animals that are endemic to the Caribbean. Hummingbirds are no exception and, although there are not that many species compared with neighboring countries to the south and west, 17 of the 22 recorded (two of which are vagrants) are endemic to the Caribbean. In fact, a few of these Caribbean hummingbirds occur on just one island or small chain of islands: this represents the highest level of endemism in the Americas.

Across the archipelago as a whole there is a wide diversity of habitats, from mangroves and coastal scrub to dense rainforest in the mountains. While many of the endemic hummingbirds favor the mountainous regions, others are most commonly encountered in the tropical gardens and breezy, open woodlands of the lowlands. One important factor that has a profound effect on the distribution of hummingbirds, and indeed of other species in the Caribbean, is the hurricane season, which generally lasts from August to October. A major hurricane can result in habitats being destroyed and hummingbird populations being decimated, often taking years to recover. In a few cases a species has been wiped out altogether from a particular island, or has disappeared for a considerable period of time, only being able to recolonize once its favored habitat has recovered sufficiently.

Distribution of Caribbean hummingbirds

● = **Caribbean endemic**
● = breeds
◆ = non-breeding visitor
◎ = vagrant Caribbean endemic
◎ = vagrant from elsewhere

SPECIES	San Andres and Providencia	Bahamas	Turks and Caicos Islands	Cuba	Cayman Is	Jamaica	Hispaniola	Puerto Rico	Virgin Islands	Anguilla	St Martin/Sint Maarten	St Barthélemy	Saba	St Eustatius	St Kitts and Nevis	Antigua and Barbuda	Montserrat	Guadeloupe	Dominica	Martinique	St Lucia	St Vincent and the Grenadines	Grenada	Barbados	No. of islands/island groups
White-necked Jacobin																							◎*		1
Rufous-breasted Hermit																							●		1
Ruby-topaz Hummingbird																							◎		1
Green-breasted Mango	●																								1
Hispaniolan Mango							●																		1
Puerto Rican Mango								●	●																2
Green Mango								●																	1
Jamaican Mango						●																			1
Green-throated Carib								●	●	●	●	●	●	●	●	●	●	●	●	●	●	●	●	●	17
Purple-throated Carib								◎	◎	◎	◎	●	●	●	●	●	●	●	●	●	●	●	●	◎	17
Cuban Emerald		●		●																					2
Hispaniolan Emerald							●																		1
Puerto Rican Emerald								●																	1
Blue-headed Hummingbird																			●	●					2
Antillean Crested Hummingbird								●	●	●	●	●	●	●	●	●	●	●	●	●	●	●	●	●	17
Red-billed Streamertail						●																			1
Black-billed Streamertail						●																			1
Bahama Hummingbird		●	●																						2
Lyre-tailed Hummingbird		●																							1
Vervain Hummingbird						●	●	◎																	3
Bee Hummingbird			◎	●																					2
Ruby-throated Hummingbird		◆		◆	◆		◎	◎																	5
Caribbean endemics (breeding)	0	3	1	2	0	4	3	5	3	2	2	3	3	3	3	3	3	3	4	4	3	3	3	2	
Total number of species recorded	1	4	2	3	1	4	4	8	4	3	3	3	3	3	3	3	3	3	4	4	3	3	6	3	

* Included on the basis of one specimen from Cariacou, which is one of the 'transboundary' Grenadines, but politically part of Grenada.

Source: Christopher J. Sharpe (see *page 279*)

Caribbean Endemic Hummingbirds—Cuba

ABOVE The **Bee Hummingbird** *Mellisuga helenae* (*left*), which is endemic to Cuba, gets its name from the curious bumblebee-like sound made by its rapidly beating wings. It is the smallest bird in the world and lives alongside the more widespread **Cuban Emerald** *Chlorostilbon ricordii* (*right*), which is almost twice its size. | Cuba

BELOW Dry forest with abundant vine tangles that is inhabited by both Bee Hummingbird and Cuban Emerald. | Matanzas, Cuba

Caribbean Endemic Hummingbirds—Jamaica

Four of the hummingbirds that are endemic to the Caribbean occur on Jamaica. In addition to the two streamertails *opposite*, two other species breed on Jamaica: the smaller **Jamaican Mango** *Anthracothorax mango* (*below* | Jamaica), which is the only hummingbird on the island that feeds from cactus flowers, and the even smaller **Vervain Hummingbird** *Mellisuga minima* (*right* | Jamaica), which is found also on Hispaniola.

FACING PAGE The most eye-catching Jamaican endemics are the aptly named **Red-billed Streamertail** *Trochilus polytmus* (*top* | Jamaica), which is present across much of the island, and the **Black-billed Streamertail** *Trochilus scitulus* (*bottom* | Jamaica) of the extreme east. Although sometimes treated as a single species, simply called Streamertail, the two differ in their calls and displays, as well as bill color.

Caribbean Endemic Hummingbirds—Other Islands

The **Purple-throated Carib** *Eulampis jugularis* (*right* | Martinique) is found only on islands in the Lesser Antilles, where the **Green-throated Carib** *Eulampis holosericeus* (*below* | Martinique) also occurs (although the range of the latter species extends farther west, to Puerto Rico). Although the two species are similar in structure and their distributions overlap, the Purple-throated Carib is generally encountered in forest at high altitudes and the Green-throated Carib at lower elevations down to sea level.

FACING PAGE **Antillean Crested Hummingbird** *Orthorhyncus cristatus* (*above*) is one of the most widespread Caribbean endemics, ranging from Puerto Rico in the west to Grenada and Barbados in the southeast, occurring in woodland, scrub and gardens from sea level to 500 m (1,600 ft). | Martinique; **Hispaniolan Mango** *Anthracothorax dominicus* (*bottom*) is, as its name suggests, restricted to Hispaniola, where it is found in a wide range of habitats throughout Haiti and the Dominican Republic. | Dominican Republic

Mexico and Central America—Increasing Diversity

Traveling south from the USA, into Mexico and through the Central American countries of Guatemala, El Salvador, Belize, Honduras, Nicaragua and Costa Rica to Panama, hummingbird diversity gradually increases. Overall, 111 species have been recorded across these countries (including the 14 that breed in the USA). Some 60 hummingbird species occur in Mexico, whereas in Central America there are 86 (the overlap with Mexico being 35 species, or 32% of the total). It is in Mexico that hermits are at the northern end of their range, as are other groups with wide distributions, such as sabrewings, jacobins, emeralds, violet-ears, mangos, coquettes and starthroats.

Some of the hummingbirds of this region are widespread and relatively easy to find, while others are restricted to particular habitats. For example, the Green-breasted Mango ranges from forest edges at 900–1,200 m (3,000–3,900 ft) right down to the coast, and is seemingly quite at home in parks and gardens, its population possibly even increasing as a result of ongoing urbanization. In contrast, although the Stripe-tailed Hummingbird occurs throughout the region, it inhabits cool, damp highland forests at around 800–2,000 m (2,600–6,600 ft), a habitat that is rapidly being lost owing to deforestation in many areas. One of the most range-restricted and threatened species in Central America is the Mangrove Hummingbird, which is, as its name suggests, one of only a few hummingbirds that inhabit mangroves (see *page 197*).

Tropical forest | Guerrero, Mexico

Hummingbirds of Mexico and Central America

RIGHT The **Black-crested Coquette** *Lophornis helenae* ranges from eastern Mexico to Costa Rica and is fairly common in open forest and clearings. It feeds mainly by traplining (see *page 34*), visiting its favored feeding sites in a regular sequence. | Costa Rica

BELOW **Violet Sabrewing** *Campylopterus hemileucurus* is a locally common, rather large and sometimes aggressive hummingbird that occurs in subtropical or tropical moist forests and tall secondary growth at 100–2,400 m (330–7,900 ft) from Mexico to Panama, some populations migrating altitudinally at different times of the year. The males form small 'exploded' leks (see *page 88*), perching in the middle strata of forests and calling persistently. | Costa Rica

ABOVE LEFT The **Cinnamon Hummingbird** *Amazilia rutila* is found in subtropical or tropical dry and moist lowland forest and dry shrubland along the Pacific coast from northwest Mexico to Costa Rica, and also around the Atlantic coast of Mexico's Yucatán Peninsula. When defending its flower-rich territories it is often a noisy and rather obvious hummingbird. | Costa Rica

ABOVE RIGHT The **Stripe-tailed Hummingbird** *Eupherusa eximia* ranges from the Gulf slope of southeastern Mexico to central Panama, favoring subtropical moist forests and adjacent clearings in the highlands. It has rufous in the wings that shows as a distinct patch at rest. | Costa Rica

BELOW **Green-breasted Mango** *Anthracothorax prevostii* ranges from southern Mexico across much of Central America to western Panama (and also occurs in northeast Colombia and northern Venezuela), occurring primarily in the lowlands up to about 1,200 m (3,900 ft). Although similar in shape and structure, the male (*left*) and female (*right*) have strikingly different plumages. Photos: Costa Rica

FACING PAGE **Long-billed Hermit** *Phaethornis longirostris* is one of the most widespread of the eight species of hermit that occur in Mexico or Central America. It is present in habitats ranging from coastal scrub and regenerating secondary-growth woodland to pristine humid forest. | Costa Rica

ABOVE The **Snowcap** *Microchera albocoronata* occurs only in Central America, its range extending from eastern Honduras and along the Caribbean coast to western Panama. It inhabits the canopy and edges of wet forest and is an altitudinal migrant, occurring from 300 m to 1,600 m (1,000–5,300 ft) depending on the time of year. | Costa Rica

BELOW **Coppery-headed Emerald** *Elvira cupreiceps* is a Costa Rican endemic, restricted to the western half of the country, where it inhabits wet forest and is patchily distributed at 700–1,500 m (2,300–4,900 ft). | Costa Rica

ABOVE LEFT The aptly named **Mexican Sheartail** *Doricha eliza* is found only in Mexico, where it is restricted to a narrow fringe of coastal vegetation along the northern tip of the Yucatán Peninsula and a small area in central Veracruz. With an estimated population of fewer than 10,000 mature individuals, which is declining owing to habitat loss, the species is considered to be Near Threatened (see *page 191*). | Mexico

ABOVE RIGHT One of the most beautiful hummingbirds of all, the **Fiery-throated Hummingbird** *Panterpe insignis*, is restricted to the highlands of Costa Rica and western Panama, occurring mainly at 1,600–3,200 m (5,200–10,500 ft). | Costa Rica

BELOW All six species of mountain-gem are found in Mexico or Central America, although some have rather restricted ranges (see maps on *page 271*). The species shown are **Purple-throated Mountain-gem** *Lampornis calolaemus* (*left* | Costa Rica) and **Green-throated Mountain-gem** *Lampornis viridipallens* (*right* | Guatemala), both of which are most frequent in highland rainforests.

South America—Incredible Diversity

Mainland South America is undoubtedly the powerhouse of hummingbird diversification, with an astonishing 279 species recorded—over three-quarters of the world's total. Of these, just 38 (14%) are shared with Central America and five have been recorded on Caribbean islands (two of these breeding, and three vagrants). Although no single species occurs throughout South America, hummingbirds taken as a group occupy nearly every habitat from sea level to the highest plateaus and mountain peaks, and can be found from the hot and humid Caribbean coast of Colombia and Venezuela to the cold and windswept tip of Tierra del Fuego. Indeed, hummingbirds have evolved to live virtually anywhere there are flowering plants, some being able to thrive even near the snowline at 5,000 m (16,400 ft) above sea level.

Fragile yet resilient: hummingbirds of the Andes

Within South America there is a clear bias in hummingbird diversity towards the equator, and there are more species in the west than in the east. It is in the tropical northern Andes, however, that hummingbird diversity reaches its peak: Colombia alone hosts 162 breeding species (this compares with 58 in neighboring Panama), and Ecuador, which is just a quarter the area of Colombia, is home to an extraordinary 136 species. The reason for the tremendous diversity of hummingbirds in these countries can be summarized in two words: The Andes.

The continuing slow but relentless uplift in altitude of the soaring mountain ranges of the Andes has had a profound effect on landform and associated habitats. As the Andes formed, the range of different habitats gradually increased and over millions of years flowering plants and hummingbirds, so inextricably linked, evolved in parallel to fill all the available niches. Although the majority of hummingbirds live in the tropical regions of the Americas, many are far from being fragile, delicate species of warm, sunny lowlands. Perhaps unexpectedly, the greatest diversity of hummingbirds is to be found where the slopes of the high Andes sweep up from lush, mist-shrouded cloud forests to stunted elfin forests and exposed, high-altitude grasslands.

Altitudinal distribution of hummingbirds

Hummingbirds occur at all altitudes from sea level to the highest mountains, sometimes even above 5,000 m (16,400 ft). This altitudinal range encompasses a wide range of habitat zones, as illustrated by the images of the Andean habitats *opposite*.

The graph *right* summarizes the altitudinal ranges of all the hummingbirds in the Americas, indicating that maximum diversity (more than 50%) of species occurs at around 1,000 m (3,300 ft). The maximum diversity of hermits, however, is found at around 400–500 m (1,300–1,600 ft), reflecting their preference for lowland forests. This broad pattern of distribution is particularly evident throughout Central and South America.

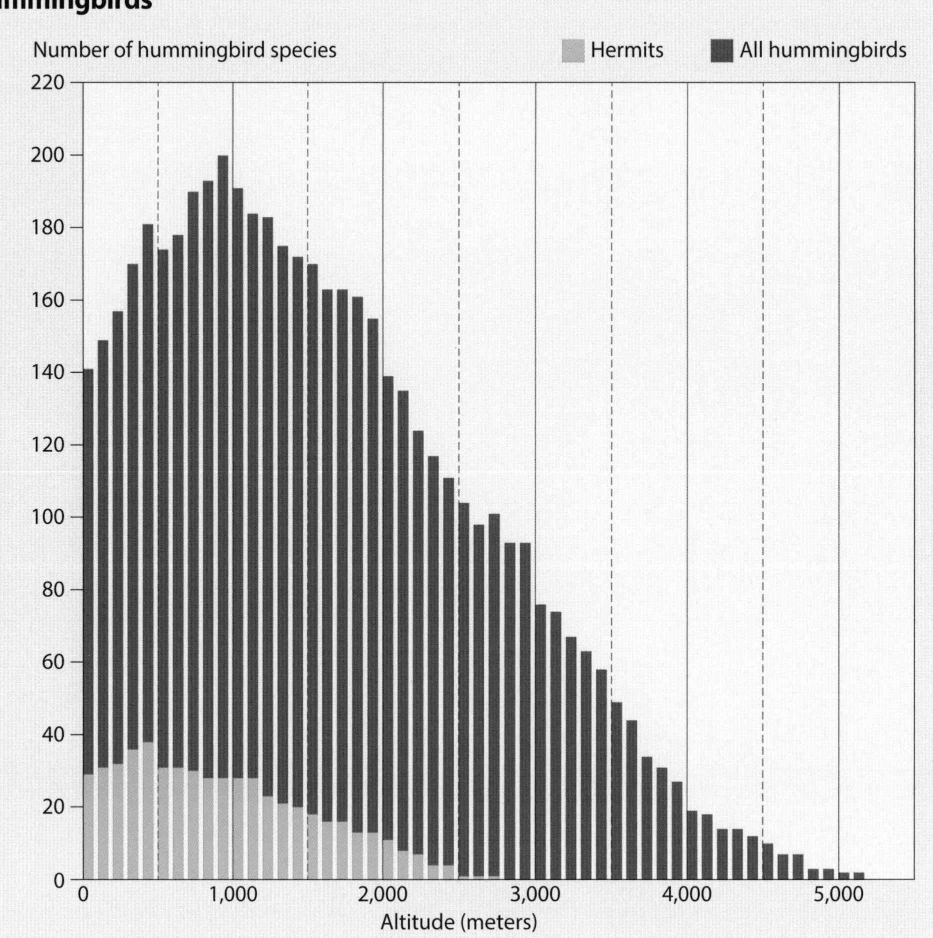

Habitats of the Andes

The Andes rise from near sea level to volcanic peaks of well over 6,000 m (20,000 ft) and have an incredibly varied topography. This has led to the development of a very wide range habitats, each of which supports its own particular assemblage of plants and associated animals, including hummingbirds. The broad habitat types shown here are explored further in this section, illustrated with photographs of many of the remarkable hummingbirds that they support. Although often not apparent at first glance, forest types in particular change markedly as you descend the altitudinal gradient.

High Andean peaks [up to 6,959 m (22,831 ft)] | Peru

Páramo [3,000–4,900 m (9,800–16,100 ft)] | Colombia

Elfin forest [3,200–3,800 m (10,500–12,400 ft)] | Ecuador

Cloud forest [2,200–3,500 m (7,200–11,200 ft)] | Peru

Fragmented foothill forest [0–2,200 m (0–7,200 ft)] | Ecuador

Coastal Desert [0–2,400 m (0–7,900 ft)] | Chile

Temperate rainforest [0–2,400 m (0–7,900 ft)] | Chile

The Andes Effect

The high peaks and precipitous, deep valleys of many of the Andean mountain ranges have for long been a very effective physical barrier to the dispersal of species, and hummingbirds are no exception. As the mountains grew ever higher, forced upwards by tectonic activity, widespread populations became fragmented, and, in many instances, were eventually restricted to remarkably small geographic areas. Gradually, these isolated populations evolved their own unique set of characteristics and became sufficiently distinct to be recognized as a species in their own right—a process that is still continuing today.

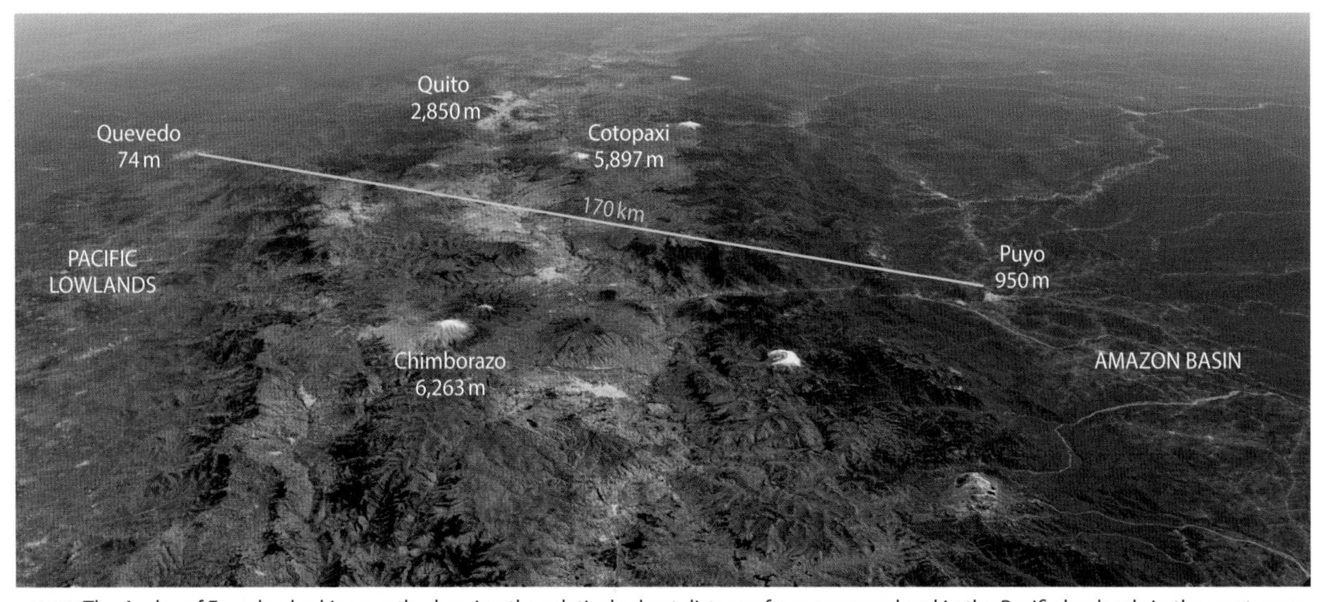

ABOVE The Andes of Ecuador, looking north, showing the relatively short distance from near sea level in the Pacific lowlands in the west, over mountain ridges that are generally more than 4,000 m (13,100 ft) in elevation, and down to less than 1,000 m (3,200 ft) in the Amazon Basin to the east—a formidable topography that is a considerable obstacle to the movement of many species. | Source: Google Earth

BELOW Steep-sided Andean valleys shrouded in cloud forest, very different from the harsh, open *páramo* above 3,000 m (9,800 ft). | Ecuador

The **Booted Racket-tail** *Ocreatus underwoodii* ranges from Venezuela through Colombia, Ecuador and Peru to Bolivia and is a typical hummingbird of the cloud-forest zone. Although this species has a particularly wide altitudinal tolerance, ranging from 850 m to 3,000 m (2,800 to 9,800 ft), and occasionally even up to 4,000 m (13,100 ft), there appears to be minimal mixing between populations on the west and east slopes of the Andes, birds on the west slope (*left* | Ecuador) having white 'boots' and those on the east slope (*right* | Ecuador) having buff 'boots'—an indication, perhaps, of speciation in progress.

The two wedge-billed hummingbirds, which inhabit tropical and subtropical moist lowland and montane forests between about 500 m and 2,600 m (1,600–8,500 ft), provide a good example of how physical geography can lead to the evolution of separate species. The **Western Wedge-billed Hummingbird** *Schistes albogularis* (*left* | Colombia), known also as the White-throated Daggerbill, is found only on the western and central Andean ridges of Colombia and southwards along the west slope of the Andes in Ecuador, while the range of the **Eastern Wedge-billed Hummingbird**, or Geoffroy's Daggerbill, *Schistes geoffroyi* (*right* | Peru) extends from Venezuela and the eastern Andean ridge in Colombia southwards along the east slope of the Andes to as far south as Bolivia. The populations do not overlap and, although structurally similar, and sharing the habit of being nectar thieves, piercing the base of flowers with their specially adapted bill, the two species are quite distinct in plumage.

Santa Marta—Endemic Hotspot

Isolated locations, particularly those that support a wide range of habitats, are most likely to result in the evolution of species which are unique to that environment. The Sierra Nevada de Santa Marta, in the very north of Colombia, is a good example of such an 'endemic hotspot'. It is one of the highest coastal mountain ranges in the world, rising to 5,710 m (18,730 ft), the northern foothills dipping towards the Caribbean Sea. This isolated massif has the highest concentration of endemic birds in South America, including five species of hummingbird. Of these, the most frequently encountered is the Santa Marta Woodstar but the others, unfortunately, are rare and threatened (see *page 199*). Although **Black-backed Thornbill** *Ramphomicron dorsale* (RIGHT) and **Santa Marta Blossomcrown** *Anthocephala floriceps* (BELOW) can be seen fairly reliably in the right locations, Santa Marta Sabrewing has been recorded only once in the last 75 years and Blue-bearded Helmetcrest was rediscovered on the highest peaks as recently as 2015, having gone undetected for nearly 70 years. | Both photos: Colombia

Living on the top of the world

The highest elevations of the Andes can be a harsh and challenging environment for both plants and birds, and any hummingbird that lives here must be resilient if it is to find enough nectar and enough insects and other invertebrates to enable it to survive. The few that do are truly remarkable, and include the helmetcrests and some of the hillstars. During short periods of adverse weather, these hummingbirds can sometimes find shelter in caves and rocky hollows, but they also have the ability to go into a state of torpor in order to conserve energy (see *page 28*). Nectar does not freeze until the temperature drops to around minus 3°C (about 27°F) (the more sugar in the solution, the lower the freezing point), but if the flowers upon which the hummingbirds depend are covered in snow for a prolonged period the birds may be forced to descend to lower altitudes for a time to seek out alternative food sources—moving back up again at the earliest opportunity.

Just below the highest peaks of the Andes, at around 3,000–4,900 m (9,800–16,100 ft), there is a cold, grassland-dominated zone that in the altiplano of Peru and south through Bolivia to northern Argentina and Chile is generally dry (known as *puna*), but farther north, through Ecuador, Colombia and Venezuela, is wet and cloudy with frequent clinging fog and drizzly rain (*páramo*). Although the weather conditions can be as demanding as those at the snowline, there are nevertheless plenty of flowers, and of course where there are flowers there are hummingbirds. Only a handful of species, however, occur here, and at low density, the *páramo* tending to support a greater diversity of hummingbirds than does the even more exposed and windswept *puna*. While living at such high altitudes undoubtedly has its challenges, there are some major advantages in that there is little competition from other birds for food and breeding territories, as well as a general freedom from predators. During the peak flowering season, other species of hummingbird from lower altitudes may also visit the *puna* and *páramo* to feed on the abundant nectar sources, moving up on warm, sunny mornings before returning again as the afternoon begins to cool.

High altitude *páramo* habitat favored by **Green-bearded Helmetcrest** *Oxypogon guerinii* (see *page 140*). | Boyacá, Colombia

Hummingbirds of the High Andes

ABOVE LEFT Often perching out in the open to feed on flowers, the **Shining Sunbeam** *Aglaeactis cupripennis* is one of the most conspicuous and frequently encountered hummingbirds of shrubby areas near the treeline throughout the high Andes of Colombia, Ecuador and Peru. It typically looks rather dull, but in the right conditions a gleaming, iridescent violet, green and gold back and rump can be seen—and the reason why this species is so named becomes obvious. | Colombia

ABOVE RIGHT Hummingbirds, such as the **Green-bearded Helmetcrest** *Oxypogon guerinii*, that live at very high altitudes have a particular need to conserve precious energy and tend to feed while perched, rather than when hovering. They favor areas with an array of flowers that offer a constant source of nectar, sitting quietly in a sunny, sheltered spot at other times. | Colombia

BELOW One of the few hummingbirds that is more or less a permanent resident of the altiplano is the **Olivaceous Thornbill** *Chalcostigma olivaceum*, a species restricted to the *puna* of Peru and northern Bolivia. This is an unusual hummingbird, as it frequently perches on the ground to feed on tiny flowers, occasionally sitting upright on a small rock ready to catch a passing insect. Of all the hummingbirds, this is perhaps the one that comes closest to moving in a genuine, if shuffling, walk as it makes its way across the dense cushions of very short grassy vegetation. | Peru

ABOVE **Eastern Mountaineer** *Oreonympha nobilis* and Western Mountaineer *Oreonympha albolimbata* are found only in the high-altitude (2,500–3,900 m/8,200–12,800 ft), dry, intermontane valleys of southern Peru. Often combined as a single species known simply as Bearded Mountaineer, the two forms are now treated as separate species by BirdLife International on the basis of differences in their head coloration, particularly the color of the crown stripe on adult males: on Eastern Mountaineer (the more widespread of the two and illustrated *here*) it is glittering violet-purple, whereas on Western Mountaineer it is scaly white and green. | Peru

BELOW The **Sparkling Violet-ear** *Colibri coruscans* is a relatively common hummingbird in a range of habitats across the Andes. Some populations breed on the open *páramo* and others do so at lower altitudes, but in both cases the birds make seasonal or daily altitudinal movements to take advantage of the peak flowering season at different locations. The bird shown here is displaying. | Peru

Into the forests

Descending lower still down the altitudinal gradient of the Andes, at around 3,500 m (11,500 ft), extensive dense patches of elfin forest suddenly begin to appear. The vegetation in this zone is perfectly adapted to the almost continuously wet, foggy, windswept conditions, and the stunted trees and shrubs, generally no more than 3 m (10 ft) tall, are draped in mosses, lichens, orchids and other epiphytic plants. Despite the conditions, there are many different brightly colored flowers, especially in shrubby clearings and along woodland fringes—and, as there is plenty of nectar, there are plenty of opportunities for hummingbirds. Here the diversity of species begins to increase and we enter the realm of metaltails, pufflegs, starfrontlets, sunangels, Great Sapphirewing, and the extraordinary Sword-billed Hummingbird.

RIGHT The metaltails are a group of nine hummingbirds that are particularly characteristic of the shrubby boundary between elfin forest and *páramo*. As is evident in this image of a **Tyrian Metaltail** *Metallura tyrianthina*, they are so called because of the gleaming metallic gloss on their tail feathers. This species is the most widely distributed of the metaltails and can be found throughout the high Andes, ranging from Venezuela to Bolivia, generally occurring at an altitude of 1,500–4,200 m (4,900–13,800 ft). All the other metaltails are more localized (see *page 262*), two species being rare and threatened. | Colombia

Elfin forest | Bosque Unchog, Peru

Hummingbirds of the Elfin Forest

The large (up to 17.5 cm/7 in long) **Great Sapphirewing** *Pterophanes cyanopterus* is a moderately common species of elfin forest and forest fringes from northern Colombia through Ecuador, Peru and Bolivia. It is found from 2,600 m (8,500 ft) to the treeline at about 3,700 m (12,100 ft). | Ecuador

Sword-billed Hummingbird
Ensifera ensifera | Ecuador

Cloud forest—hummingbirds galore

Traveling a little farther down the mountainous slopes of the Andes to around 1,800 m (6,000 ft), the trees and shrubs gradually become taller and the vegetation more luxuriant, and we enter the cloud-forest zone. For part of the day at least, as the warm, moisture-laden air of the lowlands rises to meet the colder air of the high mountains, there is more or less continual cloud cover and persistent fog or rain: the temperature in these forests remains at a fairly constant 12–16°C (54–61°F). This is one of the richest habitats in the Americas in terms of both overall biological diversity and endemism. The reason for this richness is that the zone provides the perfect conditions for plants and animals—including hummingbirds—to thrive, and for new species to evolve.

More species of hummingbird can be found in cloud forest than in any other habitat type, having coevolved with the tremendous variety of nectar-rich flowers over millennia. The range of body sizes and bill lengths and shapes is quite remarkable, reflecting the countless intricate hummingbird–flower relationships. Indeed, many plants depend upon a particular species of hummingbird for pollination, as insects, which generally act as pollinators, are not particularly well suited to such cool, gloomy conditions. Perhaps surprisingly, vertebrates, rather than insects, are often the key pollinators in such forests, with small mammals and bats, as well as hummingbirds, fulfilling this role.

Cloud forest occurs in a relatively narrow belt along the length of the Andes from Venezuela to Argentina, as well as in southeast Brazil, in mountainous areas throughout Central America and southern Mexico, and at high elevations on some Caribbean islands. Wherever they are in the Americas, cloud forests are wonderful areas to visit and in which to appreciate hummingbirds in their true element. But it is in the cloud forests of Ecuador that the greatest concentration of different hummingbirds can be encountered. For much of the year a 'core' assemblage of more than 12 species can occur, each often exploiting the plentiful nectar sources in different forest strata to minimize competition. During the peak flowering season, however, these birds are joined by migrants from both lower and higher altitudes and it is not uncommon for up to 25 species of hummingbird to be present in a single location.

Cloud forest | Ecuador

Hummingbirds of the Ecuadorian and Colombian Cloud Forests

ABOVE **Brown Violet-ear** *Colibri delphinae* (*left* | Ecuador), **Brown Inca** *Coeligena wilsoni* (*right* | Ecuador); BELOW **Rufous-gaped Hillstar** *Urochroa bougueri* | Colombia; FACING PAGE **Violet-tailed Sylph** *Aglaiocercus coelestis* | Ecuador

ABOVE **Andean Emerald** *Amazilia franciae* | Ecuador; BELOW **Velvet-purple Coronet** *Boissonneaua jardini* | (*left* | Colombia), **Hoary Puffleg** *Haplophaedia lugens* (*right* | Ecuador); FACING PAGE **Green Thorntail** *Discosura conversii* | Costa Rica

ABOVE **Empress Brilliant** *Heliodoxa imperatrix* | Colombia; BELOW **Purple-bibbed Whitetip** *Urosticte benjamini* (*left* | Ecuador), **White-bellied Woodstar** *Chaetocercus mulsant* (*right* | Ecuador); FACING PAGE **Green-fronted Lancebill** *Doryfera ludovicae* (*top left* | Colombia), **Buff-tailed Coronet** *Boissonneaua flavescens* (*top right* | Ecuador), **Fawn-breasted Brilliant** *Heliodoxa rubinoides* ♀ (*bottom left* | Ecuador), **Speckled Hummingbird** *Adelomyia melanogenys* (*bottom right* | Colombia)

ABOVE **Tourmaline Sunangel** *Heliangelus exortis* (*left* | Ecuador), **Chestnut-breasted Coronet** *Boissonneaua matthewsii* (*right* | Ecuador); BELOW **Booted Racket-tail** *Ocreatus underwoodii* | Ecuador

The foothills of the tropical Andes and adjacent lowlands

Compared with the impressive collection of hummingbirds that inhabit the Andean cloud forests, the number of species that occur at lower altitudes is far fewer. On the other hand, and unlike many of their high-mountain cousins which are restricted to a relatively small area, often because of the constraints of physical geography, the hummingbirds of the lowlands tend to be much more widely distributed. This is certainly the case for those hummingbird species that inhabit the eastern foothills of the Andes, which gradually merge into the vast expanse of the Amazon Basin.

Although much of the Andes, like elsewhere in the Americas, has been affected by deforestation and other human activities, the impact in the foothill forests and adjacent lowlands has been particularly severe. This has undoubtedly had an adverse effect on the many hummingbirds that are reliant upon intact expanses of pristine habitats. But some species, it seems, have been able to take advantage of these human-induced environmental changes, especially where such changes have led to an increase in the extent of open, sunny, flower-rich 'edge' habitats. Indeed, the ranges of a few hummingbirds have actually been expanding and their populations increasing as a consequence.

Perhaps unexpectedly, given that lowland rainforests are renowned for their exceptionally high level of species richness, they support a smaller suite of hummingbirds than do many other lowland habitats. This is principally because of a lack of sufficient suitable food sources. In dense primary forest the majority of plants flower in the open, sunny conditions of the high canopy. Although these blooms provide plenty of food for other nectarivorous forest birds, very few hummingbirds venture so high. One group of hummingbirds in particular, however, has evolved to take advantage of the limited range of flowers that are to be found in the gloomy forest understory—namely, the so-called trapline feeders that follow a regular circuit, visiting their favored flowers to exploit a nectar source that is continually being replaced (see *page 34*). The hermits, especially, are predisposed to using this feeding strategy and are one of the commonest and most vocal groups of hummingbirds throughout lowland rainforests.

Tropical rainforest in the Andean foothills | Colombia

Hummingbirds of the Andean Foothills

RIGHT The *Amazilia* hummingbirds are particularly well represented in the lowlands, some species, such as the **Glittering-throated Emerald** *Amazilia fimbriata*, being widespread and able to take advantage of floriferous open areas, often around forest clearings and human habitations. | Ecuador

BELOW The **White-necked Jacobin** *Florisuga mellivora* has a very wide distribution, being found along the foothills of both the west and the east slopes of the Andes mainly below 900 m (3,000 ft) (although it has been recorded at up to 1,900 m/6,200 ft) and throughout the lowlands of much of the northern half of South America and into Central America. It appears able to thrive in areas affected by human activity, making the most of the relatively abundant flowers associated with 'edge' habitats created as a result of deforestation for agriculture and increasing urbanization. | Ecuador

TOP LEFT Dense, often impenetrable, lowland rainforests are the domain of the hermits, such as the **White-whiskered Hermit** *Phaethornis yaruqui*, which occurs to the west of the Andes from Colombia to Ecuador. Although this species is fairly common and vocal, particularly when lekking (see *page 88*), its subdued plumage blends in extremely well with the dark undergrowth. Like all hermits, it is a key pollinator of heliconias, one of the small number of conspicuous flowering plants in the forest understory. | Ecuador

TOP RIGHT Restricted to just a few known localities along the eastern Andean foothills of Peru, **Koepcke's Hermit** *Phaethornis koepckeae*, like so many of the hermits, favors the tangled understory of undisturbed rainforest. Although relatively common in some areas, it is thought to be increasingly at risk owing to ongoing deforestation. | Peru

BOTTOM LEFT The **Violet-headed Hummingbird** *Klais guimeti* is patchily distributed along the eastern slope of the Andes, inhabiting subtropical and tropical moist lowland forests. It ranges from northern Colombia to central Bolivia (and also occurs in Central America and northern Venezuela), but despite its wide distribution the population is decreasing due to habitat loss. | Costa Rica

BOTTOM RIGHT The beautiful **Golden-tailed Sapphire** *Amazilia oenone* is a generalist hummingbird that is fairly common in forest borders and clearings along the lower east Andean slopes and valleys from Venezuela to Bolivia. It is the virtually unmistakable male that gives this species its name; as with many *Amazilia* hummingbirds, the females are rather nondescript in comparison. | Peru

The arid coastal zone

Where the foothills of the Andes meet the Pacific coast of southwestern Ecuador, Peru and the far north of Chile, the environment is relatively arid and open, encompassing the tropical dry forests of the Tumbesian region, the Sechura Desert, the Peruvian coastal plain and the Atacama Desert. A particular suite of hummingbirds is restricted to this coastal zone, including the six species shown here: one, the Chilean Woodstar, has a tiny range and is very much under threat (see *page 204*).

Tumbes Hummingbird *Leucippus baeri* | Peru

Amazilia Hummingbird *Amazilia amazilia* | Ecuador

Coastal desert | Ecuador

Hummingbirds of the Arid Coastal Zone

ABOVE **Peruvian Sheartail** *Thaumastura cora* | Peru; RIGHT, TOP TO BOTTOM **Oasis Hummingbird** *Rhodopis vesper* | Chile, **Short-tailed Woodstar** *Myrmia micrura* | Peru, **Chilean Woodstar** *Eulidia yarrellii* | Chile

Southernmost South America

As the Andes sweep farther south and into Patagonia, the climate gradually becomes cooler and rainfall increases. Although there are many nectar-producing plants in Patagonia, the diversity is much less than in tropical regions and far fewer are in bloom at any one time, particularly during the austral winter. This largely explains why, despite being the second largest country in South America, Argentina is home to only 30 species of hummingbird, while mainland Chile has just eight. South of the Atacama Desert, three species occur in the high Andes of Chile, namely White-sided Hillstar, Giant Hummingbird and Green-backed Firecrown, all of which are found also in Argentina, where they are joined by the stunning Red-tailed Comet. The Green-backed Firecrown is particularly widespread in Patagonia, especially where there are expanses of temperate rainforest, and it is the only hummingbird that braves the severe conditions of Tierra del Fuego at the extreme south of South America. It is also one of only two hummingbirds that inhabit Isla Róbinson Crusoe, one of three islands in the Juan Fernández archipelago, 670 km (415 miles) out into the Pacific Ocean to the west of Chile, the other species being the highly threatened Juan Fernandez Firecrown (see *page 204*).

ABOVE LEFT The **Green-backed Firecrown** *Sephanoides sephaniodes* is the only hummingbird that breeds as far south as Tierra del Fuego. During the austral winter these individuals migrate farther north in Patagonia, where they are dependent upon Quintral *Trycterix corymbosus* (a type of mistletoe), one of the few plants in bloom in the temperate rainforests. The firecrown pollinates the plant, which in turn depends upon the Monito del Monte *Dromiciops gliroides*, a small arboreal marsupial, to eat its berries and disperse the hard seeds in its droppings—an amazingly complex relationship. | Chile

ABOVE RIGHT The **White-sided Hillstar** *Oreotrochilus leucopleurus* is the southernmost of all the hillstars. Its range extends from the *puna* grasslands at up to 4,050 m (13,300 ft) in southern Bolivia and western Argentina south into southern Chile, where it occupies high-altitude scrublands at around 1,200 m (3,900 ft). | Chile

The southernmost tip of South America | Tierra del Fuego, Chile

The spectacular **Red-tailed Comet** *Sappho sparganurus* inhabits arid to semi-arid scrub and puna between about 1,500 m and 4,000 m (4,900–13,100 ft) in Bolivia and Argentina. This hummingbird appears to be at least a partial altitudinal migrant, descending to lower elevations during the austral winter. | Bolivia

East of the Andes—the rest of South America

Although the Andes mountains are renowned for the tremendous diversity of habitats and associated hummingbirds that can be encountered, there is also a great variety of species to be found elsewhere in South America. Unlike much of the Andes, however, habitats across the eastern part of the continent, which include the vast swathes of forest in the Amazon Basin and the hot, dry, savanna-like *cerrado*, tend to be extensive and relatively homogeneous, and do not present such extreme physical barriers to the movement of species. As a consequence, many of the hummingbirds that occur here are very widespread, but nevertheless some areas do still have their own distinctive assemblage of species, a number of which are very localized and/or rare.

RIGHT The **Swallow-tailed Hummingbird** *Eupetomena macroura* is a common and conspicuous hummingbird of open lowland areas, including gardens, across much of non-Amazonian Brazil, even prospering in cities such as São Paulo and Rio de Janeiro. It occurs also in the Guianas, Bolivia and Peru. | Brazil

BELOW The **Fork-tailed Woodnymph** *Thalurania furcata* has a very wide distribution across the northern half of South America to the east of the Andes. Although, like the hermits, this hummingbird favors the interior of humid lowland forests, it will also visit clearings, plantations and gardens in rural areas, and can sometimes be seen at higher elevations. | Brazil

ABOVE LEFT The **Black-throated Mango** *Anthracothorax nigricollis* favors open habitats throughout the tropical lowlands of northern South America up to around 1,000 m (3,300 ft) (although it has been recorded at over 1,900 m/6,200 ft), its range extending as far south as northeast Argentina and north into Panama. It can often be encountered in city-center parks and gardens and is one of the easiest lowland hummingbirds to see. | Brazil

ABOVE RIGHT The **Blue-chinned Emerald** *Chlorostilbon notatus* occurs throughout the humid lowlands of much of northern South America east of the Andes, including on Trinidad, where it is common. Its range also extends along the east coast of Brazil as far south as Rio de Janeiro. This hummingbird favors forest edge, shrubby clearings and gardens and although sometimes numerous can be inconspicuous. | Trinidad

BELOW The diminutive **Amethyst Woodstar** *Calliphlox amethystina* can be found in a wide range of open, lowland habitats across northern South America east of the Andes but is often unobtrusive. As is typical of the woodstars, however, it has a distinctive way of feeding: hovering and moving slowly in a bumblebee-like manner. | Brazil

Northern and northeastern South America

The northern fringe of South America, including Trinidad and Tobago, and inland from the Caribbean coasts of Colombia and Venezuela, as well as the Atlantic coasts of Guyana, Suriname, French Guiana and northeast Brazil, contains many different types of open habitat and lowland forest. Although relatively few hummingbirds are found here, some species are particularly characteristic, notably Buffy Hummingbird, Little Hermit and Plain-bellied Emerald, and Copper-rumped Hummingbird is common across northern Venezuela and on Trinidad and Tobago. Other species, including the stunning Tufted Coquette (see *page 6*), Green-tailed Goldenthroat and Crimson Topaz (see *page 72*), also occur but have much more extensive ranges that encompass the northern parts of Amazonia.

RIGHT **Buffy Hummingbird** *Leucippus fallax* occurs along the north coast of South America from northeast Colombia to northeast Venezuela, where it inhabits dry scrub and forest edge and mangroves. | Colombia

BELOW The **Copper-rumped Hummingbird** *Amazilia tobaci* is highly territorial and very aggressive towards other birds, even chasing off species larger than itself; it is the hummingbird most likely to be encountered in many residential areas in Venezuela and Trinidad and Tobago. | Trinidad

TOP LEFT The **Little Hermit** *Phaethornis longuemareus* occurs in tropical and subtropical forests and scrubland at up to 700 m (2,300 ft) along the northeast coast of South America from the Paria Peninsula in Venezuela to French Guiana and just into Brazil, and is found also on Trinidad. | Trinidad

TOP RIGHT The **White-tailed Sabrewing** *Campylopterus ensipennis* has two disjunct populations in the Eastern Coastal Cordillera of Venezuela and occurs also on Tobago. The fortunes of this striking hummingbird on Tobago are, however, continually in the balance: it was common on the island until 1963, when its favored montane-forest habitat was devastated by Hurricane Flora. It subsequently evaded detection until the discovery of a handful of birds in 1974, and it seems now to have returned to its former abundance, with counts even exceeding those of the early 20th century. | Tobago

BOTTOM LEFT The **Plain-bellied Emerald** *Amazilia leucogaster* is a common sight in a range of habitats, including gardens and urban areas, at up to 300 m (1,000 ft) along the northeast coast of South America from northeast Venezuela to northern Brazil, with disjunct populations in central Guyana and along the east coast of Brazil. | Brazil

BOTTOM RIGHT The **Green-tailed Goldenthroat** *Polytmus theresiae* ♀ is a hummingbird of open scrubby savannas and forest edge from 100 m to 300 m (300–1,000 ft) throughout much of the Guianas and the far north of Brazil, occasionally occurring even in coastal mangroves; it also has very localized, disjunct populations in western Amazonia and its range may actually be expanding as a result of deforestation. | Colombia

Amazonia

Much of northern South America is dominated by the Amazon Basin, with its maze of rivers and vast expanses of rainforest. Although most of Amazonia is in Brazil, the region also encompasses parts of eight other countries to the north, west and south. As mentioned earlier, the hummingbirds that inhabit the lowland forests east of the Andes are often very widespread and there is not a great diversity of species. Some, however, have very restricted distributions and, owing to the remoteness of many parts of Amazonia and the difficulties of access, are little known: there is clearly still plenty to be learned about the hummingbirds of this immense wilderness.

RIGHT The beautiful **Gould's Brilliant** *Heliodoxa aurescens* (known also as Gould's Jewelfront) is essentially a hummingbird of the interior of lowland primary rainforest in the Amazon Basin, especially near streams and creeks. It can sometimes be found, however, at higher elevations of up to 1,000 m (3,300 ft), and occasionally higher still, in the eastern foothills of the northern and central Andes. | Peru

FACING PAGE **Grey-breasted Sabrewing** *Campylopterus largipennis* (which is known also as Dusky Sabrewing) (*top*) occurs in lowland forests, typically at around 100–400 m (300–1,300 ft), across much of Amazonia and the Guianas, where it appears to favor forest edges, particularly near water. | Peru; the subtly beautiful **Pale-tailed Barbthroat** *Threnetes leucurus* (*bottom*) is found throughout most of Amazonia, ranging as far north as the Guianas and east to northern Brazil. Like the hermits and sicklebills, to which it is closely related, this hummingbird inhabits forests with a dense understory and forages by traplining (see *page 34*). | Brazil

Amazonian primary forest | Brazil

Central South America

East and south of the Amazon Basin, inland from the east coast of South America, is a vast region encompassing a huge diversity of habitats ranging from high-altitude savannas and rocky plateaus to the wetlands of the Pantanal and the grasslands of the pampas. As in Amazonia, many of the species of hummingbird found in this region are widespread, but one area in particular, in the interior of Brazil, supports a genus of two distinctive species, both of which are endemic and extremely localized: the Hooded and Hyacinth Visorbearers. Although the two have similar, specialist habitat requirements, feeding in the open and typically near the ground on terrestrial bromeliads and cacti, their ranges do not overlap. The spectacular Hyacinth Visorbearer occurs primarily in *campo rupestre* (a distinctive but very restricted habitat of open, stony meadows with a unique flora at 900–2,000 m (3,000–6,500 ft)), and occasionally in gallery forest. The Hooded Visorbearer is confined to high plateaus a little farther south in the *cerrado*.

ABOVE Despite having a very large range, the **Stripe-breasted Starthroat** *Heliomaster squamosus* is a Brazilian endemic, ranging from the northeastern to south-central and southeastern parts of the country. It occurs primarily in subtropical and tropical moist lowland forests at up to 800 m (2,600 ft) but is able to persist in areas that have been degraded, sometimes being found even in gardens, and the population is therefore considered to be stable. | Brazil

Cerrado | Minas Gerais, Brazil

ABOVE The wonderfully named **Hooded Visorbearer** *Augastes lumachella* (*left*) and **Hyacinth Visorbearer** *Augastes scutatus* (*right*) are two of the 16 species of hummingbird that are endemic to Brazil, where they are restricted to the high mountains and plateaus of the interior. | Brazil

BELOW Although characteristic of the cerrado, the stunning **Horned Sungem** *Heliactin bilophus* is highly nomadic and has a wide distribution across large parts of the eastern half of Brazil and west into northern Bolivia, with isolated outposts in northern Brazil and Suriname. This is one of the few species of hummingbird that appears to be expanding its range, reflecting its ability to adapt to open habitats created as a result of human activities, such as gardens and cultivated areas. | Brazil

The Atlantic Forest—Mata Atlântica

The Atlantic Forest was once a continuous swathe, around 3,000 km (1,800 miles) long and up to 600 km (375 miles) wide, extending from near the equator southwards along the mountainous eastern coastal belt of Brazil and west into Paraguay and Argentina. Although nowhere near as high as the Andes, these mountains still rise to an impressive 2,300 m (7,500 ft) and the associated habitats are impacted in a similar way by altitude and climate. The forest types are also broadly similar to those of the Andes, ranging from lush lowland and mid-elevation montane forests through to cloud and elfin forests. As in so many parts of the world, these forests have been devastated by humans and only a tiny fraction is left. Nevertheless, the larger forest remnants and their associated habitats, particularly in the southeast of Brazil, still support one of the highest diversities of hummingbirds in South America, with around 30 regularly occurring species.

Of the 90 hummingbird species recorded in Brazil, 18 are endemic, of which ten breed in the Atlantic Forest. Eight of these are widespread and fairly easy to see in the right locations, and they are shown here and on the following four pages, together with three other species that are encountered frequently—Black Jacobin, Violet-capped Woodnymph and Versicolored Emerald—the ranges of which extend into neighboring countries. The other two Atlantic Forest endemics—Hook-billed Hermit and Long-tailed Woodnymph—are very localized and highly threatened (see *page 205*).

ABOVE **Green-crowned Plovercrest** *Stephanoxis lalandi* | Brazil; FACING PAGE **Frilled Coquette** *Lophornis magnificus* (*above* | Brazil), **Black Jacobin** *Florisuga fusca* (*below* | Brazil)

Atlantic Forest | Serra dos Órgãos, Brazil

ABOVE **Brazilian Ruby** *Clytolaema rubricauda* | Brazil; FACING PAGE **Versicolored Emerald** *Amazilia versicolor* (*top left* | Brazil), **Sombre Hummingbird** *Aphantochroa cirrochloris* (*top right* | Brazil), **Violet-capped Woodnymph** *Thalurania glaucopis* (*bottom left* | Brazil), **Festive Coquette** *Lophornis chalybeus* (*bottom right* | Brazil)

ABOVE **Minute Hermit** *Phaethornis idaliae* | Brazil; BELOW **Dusky-throated Hermit** *Phaethornis squalidus* | Brazil; FACING PAGE The **Saw-billed Hermit** *Ramphodon naevius* is unusual among hermits in that the two sexes differ noticeably in bill shape: long and straight in males (SHOWN HERE) and shorter and slightly downcurved in females (see *page 242*). This is presumed to be an adaptation to feeding on a different range of flowers. | Brazil

Epic migrations

In much of the Americas hummingbirds are present all year round. Although some species live in more extreme conditions than might be expected, they must have food available at all times to fuel their exceptional physical demands. Unless the right plants are continuously in flower, birds must move on, either some distance away geographically (such as the Green-backed Firecrown, see *page 158*) or in a seasonal or daily altitudinal migration upslope and downslope (such as the Sparkling Violetear, see *page 141*). While subtropical and tropical hummingbirds usually migrate only relatively short distances to find food, this is certainly not the case for many that breed in the USA and Canada. If they are to exploit these northern realms, they must be long-distance migrants, moving south to warmer climes to survive the winter.

The Ruby-throated Hummingbird, common in southeastern Canada and eastern USA, undertakes a remarkable migration, although precise details are still to be unraveled. This tiny bird, which measures just 7–9 cm (2.8–3.5 in), and weighs a mere 3.0–3.5 g (0.11–0.12 oz), breeds from Texas eastwards along the Gulf of Mexico and north along the Atlantic coast to Nova Scotia and Quebec, and winters chiefly in southern Mexico and Guatemala in Central America. On migration, it might be expected to follow a route along the coast or inland between the two regions: longer than a direct sea crossing but physically much less demanding. This, however, is not always the case, particularly in spring, when many birds fly directly north across the Gulf of Mexico; in the fall, the proportion making the sea crossing is probably much smaller. Although Ruby-throated Hummingbirds that are 'fully fueled' are capable of completing a flight of up to 1,000 km (600 miles) across the Gulf (as shown on the map *opposite*), to do so they need ideal atmospheric conditions (still air or a following wind)—much more likely in spring than in the fall. A 'fast dash' north across the sea in spring, followed by a leisurely return over land in autumn, seems logical and has become entrenched in popular literature, but the true extent of migrations across the Gulf remains unclear. Whether some individuals hold to one migration strategy, or the choice of route in either season is more random, influenced by prevailing weather conditions, is also uncertain.

A 20-hour flight across the Gulf does not provide any opportunity to rest or feed, so to complete the crossing a bird must first double its weight. Everyday activities can be powered by regular sugar intake, but to meet the metabolic demands of such a long flight these hummingbirds must build up sufficient fat reserves to fuel the journey. They consume an excess of nectar, which is a challenge, especially for those birds that are making their first crossing of the open sea. Older males are generally fitter and able to put on weight

The holes (*left*) made by **Yellow-bellied Sapsuckers** *Sphyrapicus varius* create sap runs which provide an important food source for those Ruby-throated Hummingbirds that arrive particularly early on their breeding grounds early in the spring. | USA

This array of images shows the nectar plants favored by Ruby-throated Hummingbirds when on migration.
CLOCKWISE FROM TOP LEFT: **Jewelweed** *Impatiens capensis*; **Turk's Cap** *Malvaviscus arboreus*; **Dwarf Buckeye** *Aesculus parviflora*; **Trumpet Creeper** *Campsis radicans*

Many **Ruby-throated Hummingbirds** *Archilochus colubris* complete a remarkable sea crossing of the Gulf of Mexico on their spring migration northwards. In the fall, although most are believed to follow a less risky and more leisurely overland route, it appears that some do cross the Gulf. Dates on the map indicate earliest arrival in spring. | USA

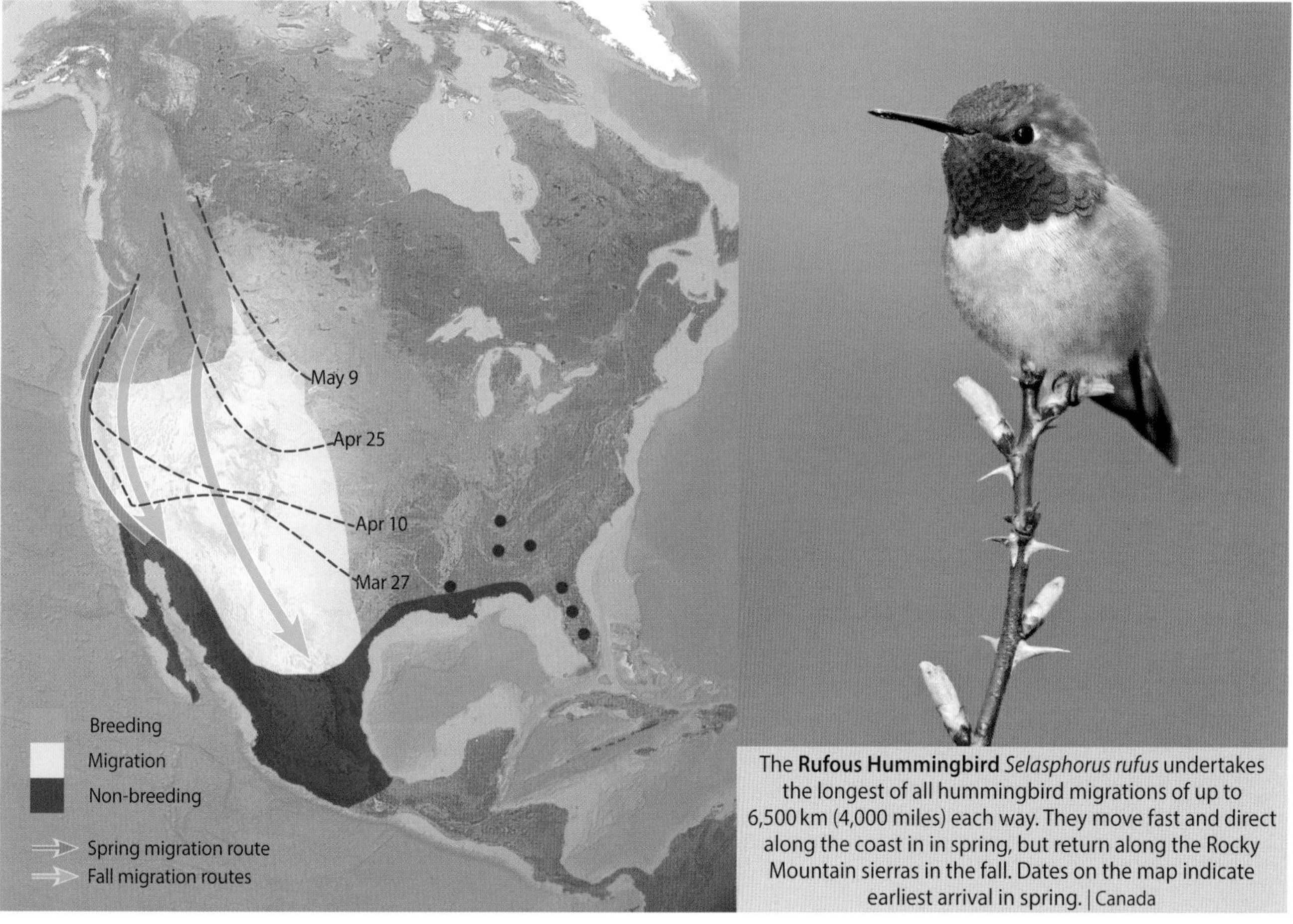

The **Rufous Hummingbird** *Selasphorus rufus* undertakes the longest of all hummingbird migrations of up to 6,500 km (4,000 miles) each way. They move fast and direct along the coast in in spring, but return along the Rocky Mountain sierras in the fall. Dates on the map indicate earliest arrival in spring. | Canada

more readily than females and younger birds, for which such a migration is certainly a tough introduction to the realities of life.

Having crossed or rounded the Gulf, Ruby-throated Hummingbirds continue directly northwards in spring, the timing apparently synchronized with the flowering of Dwarf Buckeye *Aesculus parviflora*, a favored nectar source. Older males are able to travel farther without feeding than are younger males and females, start their journey north slightly earlier, and pause for shorter periods at intermediate feeding sites. Individuals arriving particularly early on their northern breeding grounds may not find a reliable source of nectar straightaway, and depend upon the prior arrival of migrant Yellow-bellied Sapsuckers to provide an alternative source of sugary food as they open up their sap runs. All of this extraordinary effort is worthwhile in an evolutionary sense as the strongest individuals arrive on their breeding territories first and are therefore more likely to attract a mate and pass on their genes to the next generation.

In the fall, many Ruby-throated Hummingbirds take a different route and migrate southwards over land, following the Gulf coast, sometimes forming concentrations of thousands. Juveniles may need the coastline for navigation, whereas older birds can cross the sea direct. This movement coincides with a proliferation of flowers of many of their favored nectar sources, such as Jewelweed *Impatiens capensis*, Trumpet Creeper *Campsis radicans* and Turk's Cap *Malvaviscus arboreus*, which provide the necessary fuel for their onward journey. There is no obvious benefit from another risky crossing of the Gulf of Mexico at this time of year.

The Rufous Hummingbird also has a challenging migration, involving an annual journey of some 6,500 km (4,000 miles) each way between Mexico and as far north as southern Alaska—making it the northernmost breeding hummingbird. This epic flight is the longest migration of any hummingbird, and indeed the longest migration of any bird relative to its body size. During their northward migration in spring, Rufous Hummingbirds follow the Pacific coast. In the fall, they take a different route mostly along the Sierra Nevada and Rocky Mountains, often up to altitudes of around 3,000 m (9,800 ft), where there is an abundance of late-blooming alpine flowers to feed on as they go. A curving route avoids vast expanses of inhospitable desert, perhaps explaining why the Rufous Hummingbird is also the hummingbird most likely to turn up far outside its normal range. One amazing example is of an individual that was recorded in Florida, 5,680 km (3,530 miles) away from where it was banded in Alaska.

Not all of the hummingbirds that breed in the USA and Canada undertake quite such epic migrations. Anna's Hummingbird is one of the shorter-distance migrants, but its overall distribution and the extent to which it migrates have been changing. The breeding range of this species formerly extended from northwest Mexico through the Pacific states of the USA and into southern Canada. Over recent decades, however, it appears to have benefited from the introduction of irrigation schemes and the associated extensive planting of exotic flowering trees. These trees provide both a nectar source and nesting sites, and as a result this hummingbird's breeding range has gradually been expanding northwards. Such plantings are likely to be responsible for an increase in its wintering range, too, some individuals now even remaining all year as far north as southern Alaska. The effect of a changing, warming climate and a proliferation of urban hummingbird feeders may encourage a further range extension of this hummingbird.

Aside from the hummingbirds that move within North America, little is known about hummingbird migration. Many species are almost certainly resident, but others appear and disappear at locations at particular times of the year, indicating regular, annual movements. Others seem to be much more nomadic. One of the best-studied South American species is the Ruby-topaz Hummingbird. In Brazil, banding recoveries reveal movements of about 1,000 km (600 miles), some individuals departing from the southwest in April and heading northeast, returning again, often to exactly the same site, during the austral spring (October). Individuals of several other Brazilian species have been captured, transported and released up to 80 km (50 miles) away, and were able to return to the site where they were trapped with apparent ease—in the case of one individual, in as little as two-and-a-half hours!

Hummingbirds, it seems, have extremely good navigational abilities, whether traveling hundreds or thousands of miles over sea and mountains, through lowland forests to a particular patch of flowers, or from dense cloud forests up to the snowline. Although so little is known about the movements of the majority of hummingbirds, it must be the case, as with so many other bird species, that the rewards from the physical demands of migration are sufficiently great to offset the risks. Over time, the rigors of migration may also help to maintain the overall 'fitness' of a species: only those birds strong enough to complete these journeys will survive to breed and pass on their genes to succeeding generations. The propensity of some hummingbirds to migrate allows them to exploit areas that, if they were resident, would be off-limits owing to the unfavorable conditions at certain times of the year.

– o O o –

As will have become apparent, hummingbirds can be found throughout much of the Americas, and in many regions they are a very frequent sight. For this reason, and due to their remarkable attributes and dazzling plumage, they have become an integral part of human culture, not just in modern times but in ancient civilizations, too—as explored in the next chapter.

ABOVE The **Giant Hummingbird** *Patagona gigas* is a seasonal migrant in the southern parts of South America, spending the summer in the temperate areas as far south as the 44th parallel and moving north to more tropical climes during the austral winter (March–August). | Ecuador

BELOW Some **Ruby-topaz Hummingbirds** *Chrysolampis mosquitus* move as far as 1,000 km (600 miles) between breeding sites and non-breeding areas, often returning to precisely the same location year after year. Island populations are, however, largely resident. | Trinidad

Painting of hummingbirds by the German zoologist and naturalist Ernst Haeckel (1834–1919), who was renowned for his skill in merging science with art.

Hummingbirds and People:
History, Discovery and Culture

This image of the Aztec god Quetzalcoatl depicts a hummingbird feeding from a flower on the headdress.

Moctezuma II's feather headdress on display in the Museo Nacional de Antropología e Historia, Mexico City.

Hummingbirds have been revered by humankind for hundreds, if not thousands, of years, as evidenced by their strong influence on the folklore, celebrations and art of native peoples. Unfortunately, these birds' exquisite beauty has had its downside, too, as they have often been persecuted ruthlessly because of it. The use of hummingbird feathers was certainly a feature of the way of life for many indigenous cultures throughout the Americas long before the 'New World' was first 'discovered'. It was only when the first Europeans arrived, however, that the birds really began to be exploited on a massive scale, and only since that period has there been any written documentation regarding the role of hummingbirds in human culture.

Ancient myths and beliefs

Hummingbirds were a prominent feature in the spiritual life of many ancient American civilizations, but they appear to have played a particularly significant role in the mythology and culture of the Aztecs of Mexico (1438–1532 AD, known also as the Mexica). Indeed, the first Aztec king was named Huitzilihuitl, or 'hummingbird feather'. One of the most powerful Aztec gods was Huitzilopochtli, the 'hummingbird of the south' (or 'the left'), the location of the spirit world, who was linked with battles and human sacrifice, and depicted feeding on the blood of victims. Aztec warriors killed in battle were supposedly transformed into hummingbirds and after following the sun for four years returned to Earth, where they fed on nectar for eternity. The most powerful god in the Aztec world was a feathered serpent known as Quetzalcoatl, who not only wore a cape of hummingbird feathers but also had a headdress made of flowers, surrounded by feeding hummingbirds. Hummingbirds were used also as symbols for other Aztec gods, notably Xochiquetzal, the goddess of earthly pleasures, beauty and birth.

Given the important symbolism of hummingbirds to the Aztecs, it is perhaps not surprising that they were used for jewelry and ornamentation. In fact, the Aztecs are renowned for their impressive skills in creating artifacts from bird feathers of all kinds, including hugely extravagant ceremonial costumes worn by priests and royalty. There is even some suggestion that the Aztecs kept captive hummingbirds, along with many other birds, in huge aviaries to maintain a supply of feathers for such elaborate decorations. Unfortunately, when the conquistadors arrived in the early 1500s, many of these treasures were plundered and taken back to Europe, but owing to their extreme fragility hardly any have survived. One of the most famous Aztec adornments using bird plumage is what is thought to be a ceremonial headdress that possibly belonged to emperor Moctezuma II, the last of the Aztec rulers, which he reputedly presented to his Spanish conquerors in 1519. It incorporates not only 450 tail plumes of the Resplendent Quetzal, the Aztecs' sacred bird, but those of other birds, too, including hundreds from hummingbirds.

In the Nazca desert of southern Peru, within an area covering more than 440 km^2 (170 square miles), geometric lines and about 70 convoluted, figurative shapes made by the Nazca people are visible in the sand—although, remarkably, some of these can really be appreciated only from the air. Thought to have been made between roughly 2,500 and 1,500 years ago, and comprising shallow depressions and grooves about 30 cm (a foot) wide and a few centimeters/inches deep, they have been preserved only through a combination of the dry and windless conditions of the region and the nature of the underlying rock. These so-called geoglyphs, many of which depict birds, include one, just over 90 m (300 ft) in length, that clearly represents a hummingbird, the long, straight bill, pointed wings, and tail with a central 'spike' leave no doubt as to its identity. There are many theories regarding the purpose of these geoglyphs, varying from a simple religious or astronomical role to more outlandish ideas that now have little support, but suffice to say that they are still shrouded in mystery.

In North America, particularly in the dry southwest of the USA, the indigenous peoples regarded hummingbirds as 'rainbirds', perhaps because they often appeared as flowers started to bloom after a spell of rain. As a consequence, they are represented in rain-dances and appear in jewelry and decorated vessels of several native peoples. One rainbird story, from the Pueblo peoples, relates to a demon who was blinded after losing a bet with the sun and decided to set fire to the earth with hot lava. A hummingbird managed to gather clouds and put out the fire with rain and, in so doing, flew through rainbows and accumulated the bright colors in its feathers. A story from the Apache peoples tells of Wind Dancer, a deaf warrior who could sing magical songs that brought good weather and healed the sick. He married Bright Rain, after saving her from a wolf, but soon afterwards was tragically killed. There followed a bitterly cold winter, but the weather improved whenever Bright Rain went for a walk. It transpired that Wind Dancer visited Bright Rain on these walks in the form of a hummingbird, whispering sweet nothings in her ear, which brought fine weather for everyone.

There are many more recent stories, folktales, myths and legends involving hummingbirds that highlight the high esteem in which these birds were held, especially among native Americans. Several tell of how a woman fell in love with a hummingbird, but was courted by a crane. In the version told by the Cherokee (Tsalagi) peoples of the southeastern United States, the crane suggested a race around the world to settle the matter. The woman agreed, thinking that the hummingbird could fly much faster. But in the end the crane won, either because he flew in a straight line and the hummingbird flew on a zigzag path, or because the crane could fly all day and all night, while the hummingbird could fly

The Nazca hummingbird geoglyph

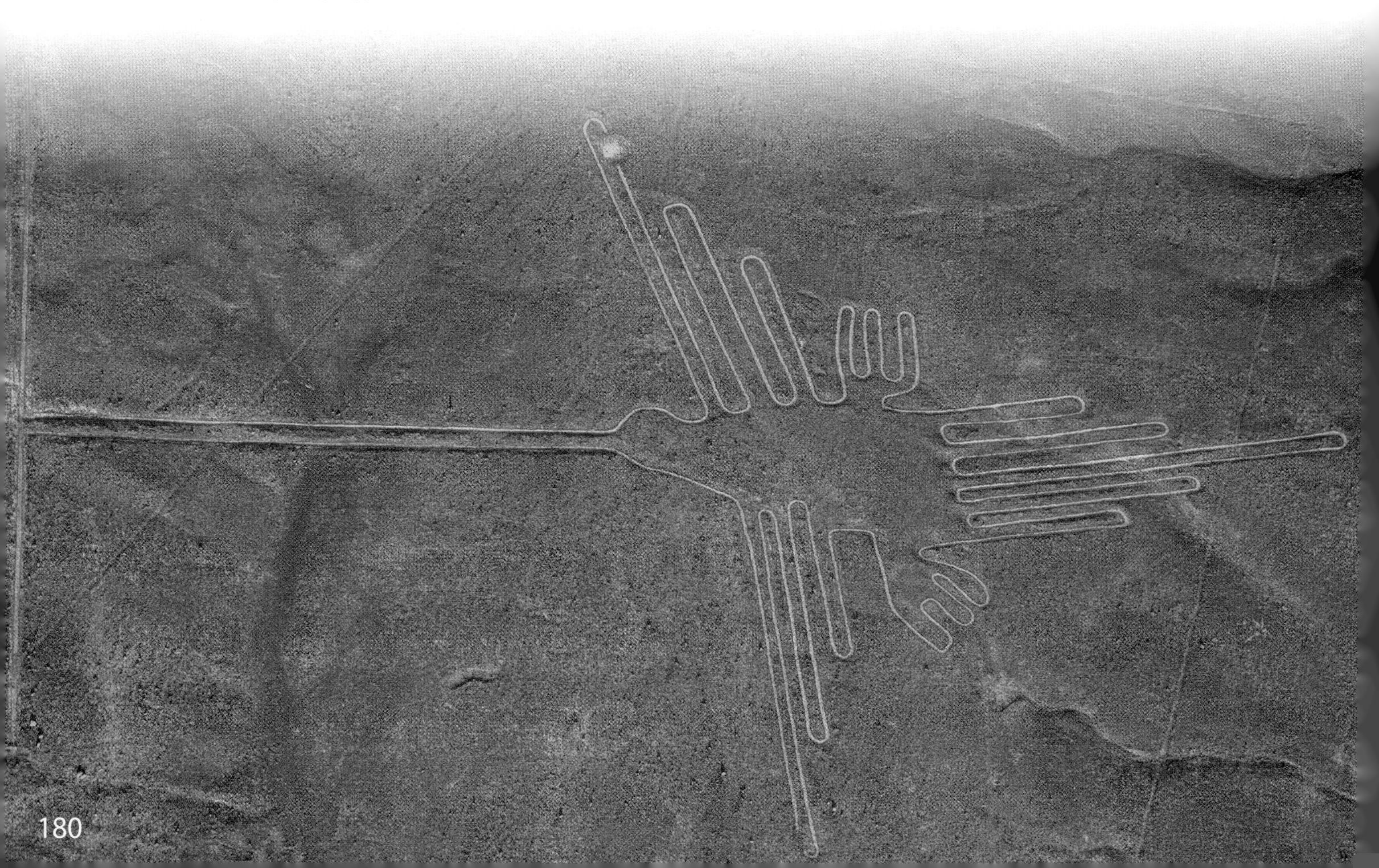

The gruesome use of dead hummingbirds in fashion accessories was all the rage during the late 19th and early 20th centuries, but is thankfully now a thing of the past (see *page 186*).

Painting: *The Humming-bird and the Crane* by Paul Bransom (1921)

The mythology of the Pueblo peoples includes more than 500 deified ancestral spirit-beings known as kachinas, one of which is a hummingbird. Provided a man in the community performs a traditional ritual while wearing a mask representing the spirit-being, it will allow itself to be seen temporarily by transforming the performer. Illustration of a hummingbird kachina by Theresa Barrow, 2019.

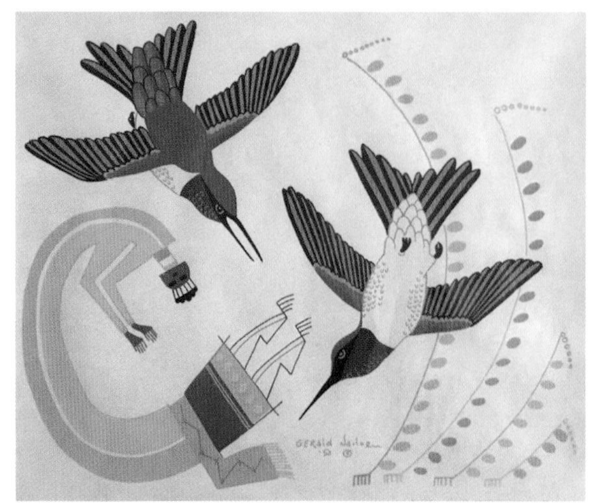

Navajo artwork by Gerald Naylor (1917–1952)

Contemporary Indigenous artwork: hummingbird by Pauline Bull, Artist, courtesy of Canadian Art Prints and Winn Devon Art Group.

only by day. Despite agreeing to marry the winner, the woman ultimately reneged on her promise as the crane was so ugly!

There are several other stories that refer to races between a hummingbird and another bird that the hummingbird should have won owing to its speed but lost because of its indecisiveness. The Hitchiti peoples (also from the southeastern United States) told of a race between a heron and a hummingbird. The heron ate large fish and the hummingbird liked small fish, but they agreed that there was enough fish for only one of them. To settle the matter they raced to a treetop, but on the way the hummingbird stopped to feed from flowers and the slow, ponderous heron won: ever since then, herons have been the fish-eaters and hummingbirds restricted to nectar.

Several legends come from the early inhabitants of the southwestern United States, notably the Navajo, Mojave and Cherokee peoples. The Navajo said that the high swoops and dives of a displaying hummingbird were attempts to see what was above the blue sky: the hummingbird found nothing but still kept on trying, just in case. Another Navajo legend recounts how the first hummingbirds were white and extremely large, destroying the flowers from which they fed in their search for nectar. To prevent this from happening the creator god reduced them in size, but unfortunately this also deprived them of their song. Considering this to be unfair, other birds requested that the creator compensate the hummingbirds by giving them the most beautiful, glittering feathers—and their wish was granted.

A Mojave story about a hummingbird is very similar to that told by Europeans about a dove: people lived in darkness underground but finally found a way to the light after sending a hummingbird up through narrow, twisting passages to the world of sunlight. A hummingbird also has the same role as a dove in another story, which relates to how it flew off and returned with a sprig of foliage, thereby providing good news that a great flood was finally subsiding.

In Cherokee legend, hummingbirds appear as 'tobacco birds', bringing tobacco to people, or smoke to shamans, who use it to purify the earth, the tobacco perhaps obtained from Caterpillar, guardian of the tobacco plant. One story tells of the fearsome Dagul'ku geese that stole all the tobacco plants and defeated attempts by people and four-footed animals to retrieve them—but a hummingbird was able to dash in undetected, pick some leaves and seeds, and return tobacco to the tribes.

Hummingbird legends also extend to the Caribbean islands. The native Taíno, in what is modern-day Puerto Rico, told the story of two young people who fell in love, to the intense dislike of their families and friends; but the two rose above all of this when the young man became a hummingbird and his bride a red flower. The early Arawak inhabitants of what is now Trinidad called the island *Iëre*, meaning 'land of the hummingbird', and revered the birds as long-dead ancestors: as a consequence, they were never harmed.

A legend from the Mayan civilization of Mexico and Central America tells of a hummingbird, called Tzunuum, who was dull-colored but proud of his spectacular feats of flying. At his wedding, however, he was concerned about his dull appearance and so other birds came to his rescue, giving him fragments of their colorful feathers as a wedding gift. Because Tzunuum was so grateful for this gift he was allowed to wear brightly colored feathers ever after—but he was almost speechless in his gratitude and this is why hummingbirds lost their songs. An alternative Mayan story is that the Great Spirit created birds, but as there were pieces of bright feathers left over he then created a pair of hummingbirds, to which the sun added its glittering light. A further Mayan legend also tells that a hummingbird is in fact the Sun in disguise, trying to court the Moon.

Examples of Lesson's paintings from his works on hummingbirds, *Histoire Naturelle des Oiseaux-Mouches* (1829): **Purple-crowned Fairy** *Heliothryx barroti* (*left*) and **Stripe-breasted Starthroat** *Heliomaster squamosus* (*right*).

The role of the hummingbird in such stories is clearly varied, yet there are particular threads that reappear over time and across large areas of the Americas. Hummingbirds clearly made an impression on people whenever and wherever they were encountered, as they still do today.

European contact

Our detailed knowledge of hummingbirds inevitably begins with a European perspective, as early explorers in the 16th and 17th centuries became obsessed with discovering all the untold excitements of a 'New World'—the Americas. Yet, despite the multitude of fascinating new peoples, cultures and treasures that confronted them, Europeans were still so astonished by hummingbirds that they took time to write about them and wonder at what manner of creatures they were seeing. Some believed that these extraordinary tiny beings, darting between flowers and fighting among themselves, might be some kind of cross between birds and insects. How could they fly like that? How could they appear dark or colorless one moment, yet brilliantly patterned the next? In around 1500, Christopher Columbus wrote in his diaries about a vast range of new experiences, but he was sufficiently struck by hummingbirds that he was compelled to mention them and raise his uncertainty about what kind of animal they might be.

Upon first contact with indigenous peoples the early explorers soon met high officials and Aztec kings, and marveled at their extravagantly decorated robes and other artifacts, many of which featured the feathers of hummingbirds—the likes of which they had never seen before. It was, however, the living birds that so often astounded them above all else. In a journal published in 1557, a mariner, Jean de Léry, traveling along the Brazilian coast mentioned a bird whose body was "*no larger than that of a hornet or a stag beetle*" and remarked that it was "*an extraordinary wonder and a masterpiece of minuteness.*" Stories about hummingbirds began to filter back to Europe: these were magical birds that, some said, stick their beak in tree trunks and die in autumn, only to be restored to life in spring. There were also rumors that hummingbirds would strike at the eyes of people with their needle-like bill, leading to a widespread belief that these tiny birds were dangerous. Hummingbirds were in fact deemed so remarkable and worthy of attention that artifacts decorated with their skins made their way to Europe, at least one apparently even being given as a gift to the Pope.

It was a long time, however, before hummingbirds became at all well known in Europe. As relatively recently as 1758, the great Swedish systematic naturalist Carl Linnaeus was able to catalog only 18 species of hummingbird (today, 371 species are recognized). In 1829, the first of three splendid volumes on the world's hummingbirds was published by the French naturalist and naval surgeon René Lesson. In these books he referred to hummingbirds as *oiseaux-mouches*, or fly birds, and the first volume included 261 lithographs and described almost 40 new species: one of these Lesson named Anna's Hummingbird after Princess Anna de Belle Massena, the wife of a French prince.

Between 1849 and 1861, British naturalist, collector and artist John Gould trumped Lesson with a five-volume set, *A Monograph of the Trochilidae, or family of humming-birds*, which featured 360 magnificent hand-colored lithographs, heightened with gold-leaf and iridescent mineral paints. This huge work was completed after his death and remains one of the greatest achievements in the history of ornithology. In it, Gould claimed to identify 400 species of hummingbird—about 30 more than are currently recognized. In the introduction, he wrote "*these wonderful works of creation... my thoughts are often directed to them in the day, and my night dreams*

An illustration from Gould's five-volume *A Monograph of the Trochilidae, or family of humming-birds* depicting **Crimson Topaz** *Topaza pella*.

An illustration from Gould's five-volume *A Monograph of the Trochilidae, or family of humming-birds* depicting **Marvelous Spatuletail** *Loddigesia mirabilis*.

have not infrequently carried me to their native forests in the distant country of America." Through his books and exhibits in England, John Gould was the first person to allow Europeans to appreciate the variety and spectacular nature of hummingbirds to the full.

Gould had been the naturalist entrusted to identify specimens brought back from the second *Beagle* expedition by Charles Darwin, and in 1832 Darwin himself had written: "*As we passed along, we were amused by watching the humming birds. I counted four species—the smallest at but a short distance precisely resembles in its habits & appearance a Sphinx. The wings moved so rapidly, that they were scarcely visible, & so remaining stationary the little bird darted its beak into the wild flowers, making an extraordinary buzzing noise at the same time, with its wings. Those that I have met with, frequent shaded & retired forests & may there be seen chasing away the rival butterfly.*" The Sphinx, in this context, is a large, long-winged moth that does indeed resemble a typical small hummingbird (see *page 10*), but Darwin had little reference material available to him and naturalists were still only just beginning to understand the lives and diversity of hummingbirds.

Throughout the 19th century, species new to science were continually being found worldwide, and collectors of birds were being sent ever more remarkable specimens. To collect hummingbirds was to share in some of the most exciting discoveries from the most exotic locations, and each new discovery to reach Europe seemed even more spectacular than the previous one. Once the word was out about the incredible beauty of these tiny, exotic birds, collectors from around Europe began to covet and seek them out for their personal exhibits. Hummingbird collections, set in expensive decorative glass cases and containing up to 100 dead birds, became popular items in stylish Victorian homes—seemingly no more reprehensible than a fine collection of pinned butterflies and moths, the survival of any of the species concerned being of little consequence.

In 1824, William Bullock, the then curator of the Natural History Museum in London, UK, who considered his hummingbird cabinet to be among his finest exhibits, wrote: "*There is not, it may safely be asserted, in all the varied works of nature in her zoological productions, any family that can bear a comparison, for singularity of form, splendour of colour, or number and variety of species, with this the smallest of the feathered creation.*"

Another cabinet, containing 326 mounted hummingbird specimens, has also been held at the Natural History Museum, Tring, UK, since the late 19th century. This cabinet represents a small part of the collection of British traveler and collector Oscar Baron, who journeyed across the Americas. Baron's hummingbirds were apparently mounted 'in the field' in lifelike postures, and the display, which represents one

Baron's hummingbird display cabinet in the Natural History Museum, Tring, UK. | Reproduced courtesy of The Trustees of the Natural History Museum, London.

hundred species of hummingbird, comprises not only skins but also nests and eggs.

Over the years, ornithological studies have often resulted in the collection of specimens and the inevitable casualties of the species being studied in the name of science. The scale of such collecting, however, pales into insignificance in comparison with the level of persecution that took place from around the end of the 19th century for a very different reason. At this time, and into the early 20th century, the desire for hummingbird skins and feathers for the millinery (women's hat) trade and related industries accounted for many millions of birds, all exported from Central and South America to the markets of Europe and North America. They were used to create what now appears to be bizarre, gruesome jewelry, including intact hummingbird heads used whole, rather like large gemstones, mounted on brooches and even as earrings (as illustrated on *page 181*). Innumerable feathers were used on clothes and ornaments; in one documented case, 8,000 birds were incorporated into a single shawl. The figures relating to the scale of the import are astounding, with individual sales sometimes amounting to tens of thousands of bird skins. For example, a single London lot comprised 37,603 South American skins, and a consignment from Brazil included 3,000 Ruby-topaz Hummingbird skins alone. In fact, between 1904 and 1911 one London auction house sold 152,000 skins—a figure that is likely to be more than the entire number of hummingbird specimens currently held by all the world's museums.

Insensitive traders and fashion houses were not the only ones responsible for the deaths of so many hummingbirds. John Gould, who produced the superb five-volume monograph on hummingbirds mentioned earlier, along with other wonderful volumes illustrating the world's birds, exhibited 1,500 stuffed hummingbirds in London's 1851 Great Exhibition (which Queen Victoria herself saw as a visitor). Gould, to be fair, also tried to bring live hummingbirds to Britain to be kept in captivity, but they died soon after their arrival: nobody really knew how to keep them, a hummingbird that arrived at London Zoo in 1905 surviving for just two weeks. All of this intense interest in hummingbirds, with such obvious appreciation of their beauty and astonishing lifestyles, perversely led to cruelty and the uncontrolled killing of the very creatures that had become so highly regarded and revered.

In the United States, the hummingbird became a symbol of freedom and a 'brave new world' following the tribulations of the 1861–65 Civil War. So great was the interest that hummingbirds were interpreted almost as a 'national obsession'. The 19th-century poet Emily Dickinson wrote about hummingbirds and even signed letters and notes from herself "Hummingbird", as if she felt that she was the embodiment of such a lively and delicate creature. Artist Martin Johnson Heade, a painter of romantic scenes including flowers and hummingbirds, was obsessed, too: he wanted to match Gould's work in Europe with a series of paintings, together with detailed annotations on the lifestyles and biology of hummingbirds, in a new publication, *The Gems of Brazil* (1863–64). Specimens provided him with the details but he needed living birds to bring life and veracity to his pictures, so he went to South America, just as the Civil War began. Heade was passionately anti-slavery and in touch with other prominent abolitionists, and although, sadly, his project failed to materialize, hummingbirds became, in one way or another, emblems of freedom, as noted by other anti-slavery activists such as Harriet Beecher Stowe. The ability of the birds to move freely on their migrations no doubt contributed to this connection between hummingbirds and freedom.

Long after the earliest adventurers and explorers had set foot in the Americas, European travelers continued to be awed by hummingbirds, some writing effusively of their observations. Even North American observers, despite having personal experience of a few species back home, were often equally affected by their encounters with hummingbirds when traveling farther south. In the mid-19th century, the American collector and artist John James Audubon wrote passionately about hummingbirds, and his reflections are familiar to hummingbird enthusiasts today. Of the mangos, he said "...[they] *are the greatest ornaments of the gardens and forests. Such in most cases is the brilliancy of their plumage, that I am unable to find apt objects of comparison unless I resort to the most brilliant gems and the richest metals.*" Nevertheless, despite this growing appreciation for hummingbirds, collecting was, unfortunately, still the unashamed norm. Audubon also wrote "*My good friend Thomas Nuttall while travelling from the Rocky Mountains toward California happened to observe on a low oak bush a Humming bird's nest on which the female was sitting. Having cautiously approached he secured the bird with his hat. The male in the meantime fluttered angrily around but as my friend had not a gun he was unable to procure it.*"

Hummingbirds in modern culture—the 20th and 21st centuries

Thankfully, with the dawn of the 20th century, attitudes towards the natural world were beginning to change. Another famous American naturalist who became enchanted by hummingbirds was Alexander Skutch (1904–2004). In 1928, Skutch arrived in Panama as a young botanist ready to study tropical fruits such as bananas. Looking out of his office window one day, he noticed a hummingbird building a nest, and his ensuing fascination with the proceedings triggered a lifetime's devotion to bird study. Skutch believed that beautiful creatures, hummingbirds included, were beautiful for a reason—simply in order to be appreciated. He wrote "*In my youth I knew the delight of watching the beauty, wonder and mystery of the natural world unfold before my developing mind; as when one who has climbed to a mountain-top in the night watches the dawn reveal the glorious panorama spread out before him. I have never outgrown that delight, and I hope that I never shall.*"

Skutch considered birds to be sentient beings, aware of what they are doing, rather than going through life unconscious of their actions as is so often assumed. He would go on to write an amazing number of books and papers about many species of bird, including a splendid book on the life of the hummingbird (see *Further Reading, page 277*). Skutch felt that he had a duty to share his findings and to seek out wondrous things that had not yet been discovered or properly appreciated: to maximize what he called the 'latent values' of everything beautiful and wonderful in the world. He recalled: "*Every beauty lurking unseen and unadmired, every lovable creature living obscurely, every absorbing fact remaining to be discovered, deep in remote forests or high on forbidding peaks, seemed a challenge to seek it out and make it known*." This was the challenge that he willingly accepted for the duration of his lengthy and accomplished ornithological career—and it all began with a fascination with hummingbirds. Modern-day students of hummingbirds, with so much technology at their disposal, may use different techniques from those of Skutch, but, like him, they surely are all equally entranced by the simple intrinsic beauty of the subjects of their study.

As time has moved on, stories and legends associated with hummingbirds have gradually diminished. In some parts of the Americas, however, the hummingbirds' power continues to be celebrated to this day in folklore and superstitions, albeit sometimes in strange and nefarious ways. Hummingbirds feature in the indigenous Taurepán (Pemón) culture of the Gran Sabana in southeast Venezuela, being considered sacred and believed to protect homes: newly constructed houses are inaugurated by the performance of hummingbird-themed dance rites, during which a particular kind of music called *tukuy* is played. Another example comes from Colombia, where indigenous music is based largely on bird song; traditional hummingbird-themed music is never played in the same way twice, seeming to have a beginning but no distinct end. Sadly, though, hummingbirds are still being persecuted owing to misguided beliefs. In Mexico, for example, hummingbirds are killed illegally for use as gruesome love charms, called *chuparosas*, which increasingly are being smuggled into the USA to feed a growing black market. Hummingbirds are also considered 'messengers from the heavens' and 'road-openers' for travelers, and dead hummingbirds have been found in shrines used by drug cartels to pray to patron saints for safe passage, good luck and protection from the police.

The relationship between people and hummingbirds remains complex, but the birds have always been, and continue to be, used as potent symbols in many aspects of everyday life. Hummingbirds feature in the graphic designs used by many companies to promote their products, ranging from hi-tech materials and investment companies to musical instruments, and from computer search engines to bakeries, bicycles and airlines. Disney has featured hummingbirds in several of its films, including the character Flit in the 1995 animated feature film *Pocahontas*. Despite their small size and lack of physical power, hummingbirds remain popular as tattoos, as they are regarded as symbols of beauty and intelligence.

The local names for hummingbirds vary throughout the Americas and yet all seem to relate to their flight, the sound they make when flying or their feeding behavior. For example, in Brazil hummingbirds are called

The cartoon character (*left*) is Flit, from Disney's 1995 animated feature film *Pocahontas*. Hummingbirds' aerial prowess naturally attracts airlines and other aerial services when they look for a suitable company logo.

Back in 1962, when a new coat of arms was being designed for Trinidad and Tobago, two stylized 'golden' hummingbirds were included on the central shield, together with a Scarlet Ibis and Rufous-vented Chachalaca (or Cocrico), all birds that were regarded as being particularly representative of the islands

beija-flores (flower-kissers); in Spanish they are known as *chupaflores* (flower-suckers) or *picaflores* (flower-nibblers), as well as the more regular name *colibríes* for hummingbirds in general. In Cuba, the larger Cuban Emerald is known as the *zunzún* while the smaller Bee Hummingbird is the *zunzuncito*, both names apparently derived from the word *zumbar*, which translates broadly as "*buzz-buzz*".

– o O o –

In the modern world, there really is no good reason to kill or harm hummingbirds. In fact, there is every reason for us to appreciate them for their own sake—wild, free and so vividly alive—and to do what we can to safeguard their future.

FACING PAGE **Chestnut-breasted Coronet** *Boissonneaua matthewsii* | Ecuador

Conservation: Hummingbirds Under Threat

The threats that hummingbirds face

The fortunes of hummingbirds, and the habitats upon which they depend, have been in the balance ever since their first contact with humans. Paradoxically, it was their spectacular beauty that at one time proved to be their curse, but thankfully one of the major threats that hummingbirds faced in the past has largely ceased. That threat was the indiscriminate and uncontrolled trade in hummingbirds in the 19th century, particularly of their skins, the scale of which was quite astounding, as explained in the previous chapter, *Hummingbirds and People* (*page 179*). Even as early as the 1850s it was already becoming clear that populations of some hummingbirds were declining and that some species were disappearing altogether from large areas. Careful checking of hummingbird artifacts from around this time also suggests that hunting and trapping may even have led to some species being totally eradicated (although the problems of identification of species from single skins are considerable, and 'one-off' hybrids may account for some of the specimens of uncertain identity). Fortunately, the trade in hummingbird skins came to an end during the early years of the 20th century.

Nowadays, hummingbirds, unlike with many other groups of birds, are for the most part no longer subjected to direct human persecution, as they tend to be viewed as entirely benign: they are neither livestock-predators nor agricultural pests and are too small to constitute a worthwhile meal. Indeed, given their aesthetic beauty, long cultural association with humans, and role as pollinators, they are generally welcomed and cherished. Even so, the problems associated with their beauty have not been eliminated entirely. Hummingbirds are still being trafficked illegally for the pet trade (all species are covered under Appendix II of the Convention on International Trade in Endangered Species (CITES), which prohibits export without a specific license) and smugglers carrying anaesthetized birds within their clothing are sometimes apprehended at international airports.

Although the trade in hummingbirds may still possibly be having an effect on some species, the threats that they currently face are mostly indirect and unintended and can largely be attributed to human-induced habitat loss and degradation. Natural phenomena, too, such as drought, wildfires and hurricanes, can also have an adverse impact on hummingbirds, the last of these being a problem primarily for isolated populations on islands (see *Biogeography and Biodiversity: The Hummingbirds' Realm, page 122*). There is, of course, one other threat that could increasingly have an impact on hummingbirds, and that is climate change, which may affect the nature of their habitat and the incidence and severity of extreme weather events.

Assessing conservation priorities

Increasingly, it seems, action needs to be taken to prevent species from becoming extinct, but before conservation efforts can be targeted effectively it is important to identify in a robust and consistent manner those that are most at risk. This process really began only in 1964, with the establishment of the IUCN (International Union for Conservation of Nature) *Red List of Threatened Species*. This list is the standard source of information on the conservation status of the world's species, and is the product of collaboration by a global network of many thousands of volunteers who provide expertise and information on the changing fortunes of the nearly 135,000 species assessed so far. As a relatively well-known taxonomic group, birds are fortunate to have been covered in their entirety. The evaluations for birds are coordinated by BirdLife International, which requests and collates information on all species, applies rigorous IUCN criteria to determine their conservation status, publishes annual updates on the conservation status of all the world's birds, and encourages proposals for reappraisal.

Species are assigned to categories on the IUCN Red List on the basis of criteria with clear numeric thresholds relating to five measurable parameters: population size, range size, distribution, rate of population decline, and rate of range contraction. Every four to six years the entire Red List is subject to a full review, but even in intervening years the category assigned to a species may change, as a result either of real changes in the severity of the threat(s) affecting a species or of improved knowledge of these threat(s) and how a species is likely to respond to them.

The overall state of the world's birds makes for rather sobering reading. Of the 11,003 extant species recognized by BirdLife International at the end of 2021, as many as 2,485—approaching a quarter—are of conservation concern. In total, 1,481—an eighth of all species—are considered to be at risk of extinction and are therefore placed in one of the three highest threat categories ('threatened' on the chart on *page 192*). Compared with many families of birds, hummingbirds are not faring

FACING PAGE **Sapphire-bellied Hummingbird** *Amazilia lilliae* (Endangered) | Colombia

quite so badly, with 62 species (one sixth) being of conservation concern, of which 39 (one in nine species) are considered to be Globally Threatened: nine are Critically Endangered, 17 Endangered and 13 Vulnerable. The remaining 23 species that are of conservation concern do not quite meet the criteria to be considered Globally Threatened but could easily do so in the near future, and are therefore categorized as Near Threatened.

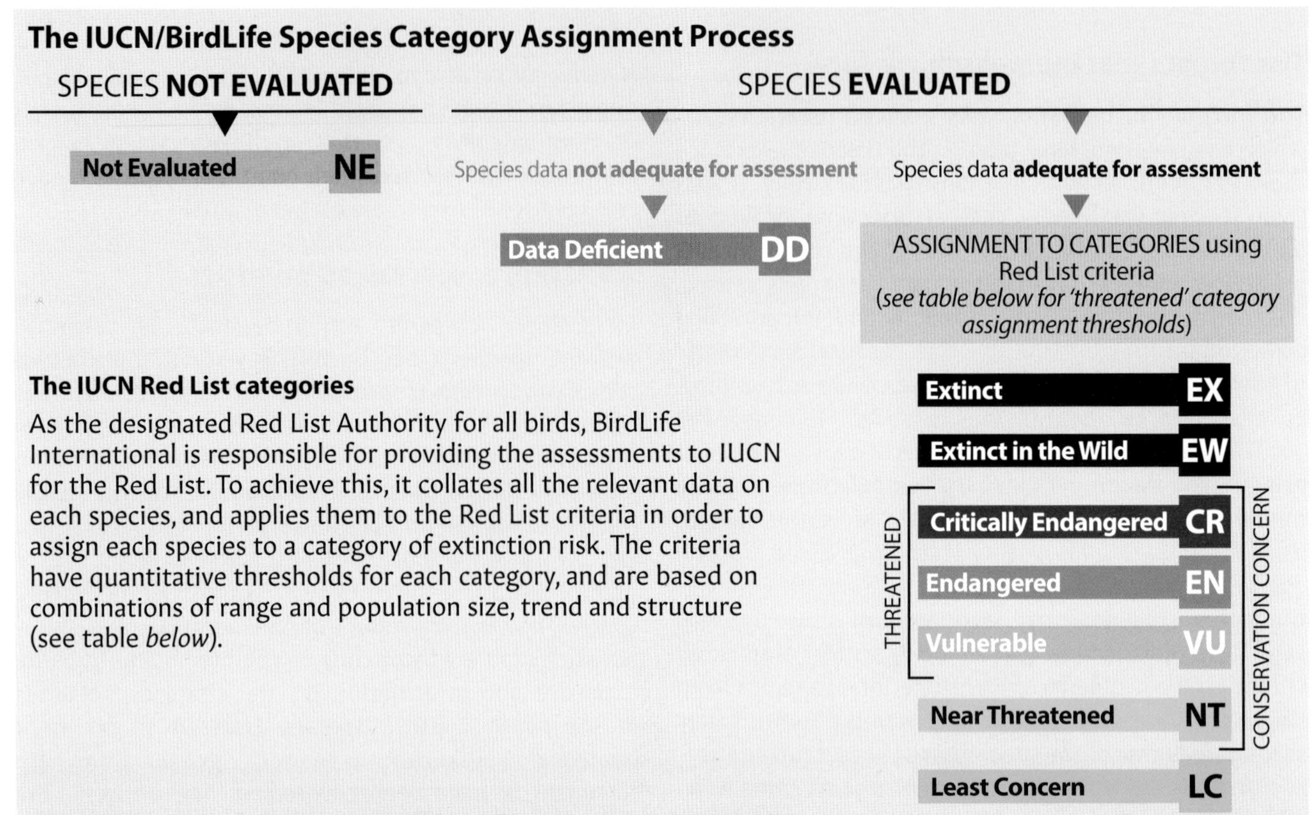

The IUCN Red List categories

As the designated Red List Authority for all birds, BirdLife International is responsible for providing the assessments to IUCN for the Red List. To achieve this, it collates all the relevant data on each species, and applies them to the Red List criteria in order to assign each species to a category of extinction risk. The criteria have quantitative thresholds for each category, and are based on combinations of range and population size, trend and structure (see table *below*).

Simplified overview of thresholds for the IUCN Red List criteria

The IUCN Red List is updated annually, with a comprehensive review of all bird species undertaken every four to six years.

Criterion		Critically Endangered (CR)	Endangered (EN)	Vulnerable (VU)	Qualifiers and notes
A1:	Reduction in population size	≥ 90%	≥ 70%	≥50%	Over 10 years / 3 generations in the past, where causes are reversible, understood and have ceased.
A2-4:	Reduction in population size	≥ 80%	≥ 50%	≥30%	Over 10 years / 3 generations in past, future or combination.
B1:	Small range (extent of occurrence)	<100 km²	<5,000 km²	<20,000 km²	Plus two of (a) severe fragmentation / few localities (1, ≤5, ≤10), (b) continuing decline, (c) extreme fluctuation.
B2:	Small range (area of occupancy)	<10 km²	<500 km²	<2,000 km²	Plus two of (a) severe fragmentation / few localities (1, ≤5, ≤10), (b) continuing decline, (c) extreme fluctuation.
C:	Small and declining population	<250	<2,500	<10,000	Mature individuals. Continuing decline either (1) over specified rates & time periods or (2) with (a) specified population structure or (b) extreme fluctuation.
D1:	Very small population	<50	<250	<1,000	Mature individuals.
D2:	Very small range	n/a	n/a	<20 km² or ≤5 locations	Capable of becoming CR or EX within a very short time.
E:	Quantitative analysis	≥50% in 10 years / 3 generations	≥20% in 20 years / 5 generations	≥10% in 100 years	Estimated extinction risk using quantitative models *e.g.* population viability analyses.

Summary of the conservation status and population trends for the world's hummingbirds in comparison with all the birds of the world (as at December 2021)

Category	Hummingbirds		All birds	
EX	2	0.54%	159	1.42%
EW	—	—	5	0.04%
CR	**9**	**2.43%**	**225**	**2.02%**
EN	**17**	**4.58%**	**447**	**4.00%**
VU	**13**	**3.50%**	**773**	**6.93%**
NT	**23**	**6.20%**	**1,010**	**9.05%**
LC	306	82.48%	8,493	76.09%
DD	1	0.27%	50	0.45%
Total	**371**		**11,162**	

Category	Hummingbirds		All birds	
Increasing	11	2.98%	664	6.04%
Stable	88	23.85%	4,240	38.55%
Decreasing	**178**	**48.24%**	**5,393**	**49.04%**
Unknown	92	24.93%	701	6.37%
Total *	**369**		**10,998**	

* Figures exclude species that are EX or EW.

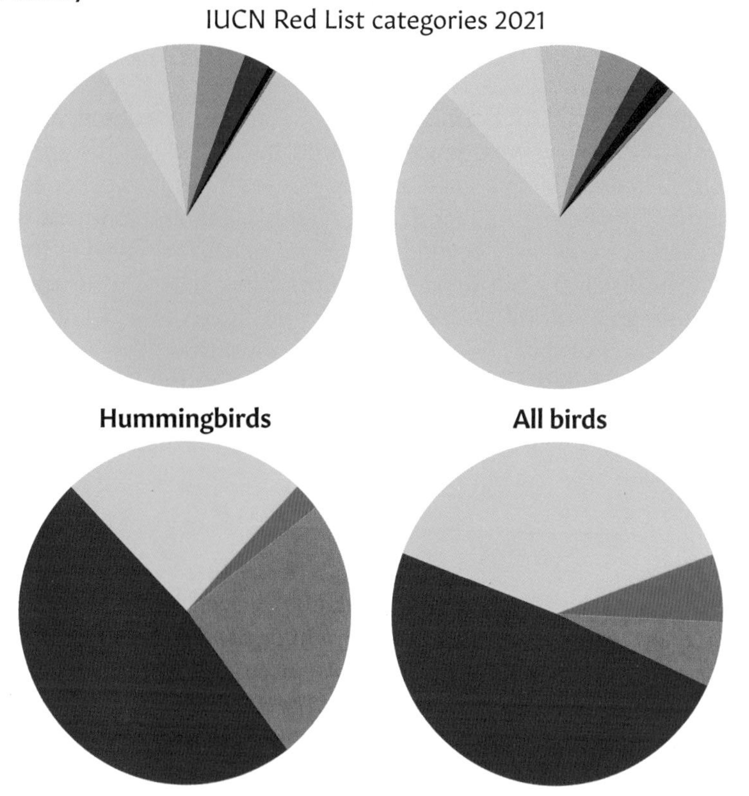

Summary of the distribution of hummingbirds of conservation concern

The number of Globally Threatened and Near Threatened hummingbird species that breed in each country (including associated islands).

	Species recorded (excluding vagrants)	CR	EN	VU	Globally Threatened [shared]	NT	Conservation concern
NORTH AMERICA							
Canada	5	—	—	—	—	1	1
USA	14	—	—	—	—	1	1
Mexico (including the islands of Cozumel and Tres Marías)	60	1	1	1	3	4	7
THE CARIBBEAN							
All islands (excluding Trinidad and Tobago)	20	—	—	—	—	1	1
CENTRAL AMERICA							
Honduras	41	—	—	1	1	—	1
Costa Rica	53	1	1	—	2	—	2
Panama	58	—	1	—	1	2	3
SOUTH AMERICA							
Colombia	161	3	6 [2]	6 [2]	15 [4]	8	23
Ecuador	136	2	3 [1]	3 [2]	8 [3]	6	14
Peru	132	—	3 [1]	2 [1]	5 [2]	6	11
Bolivia	79	—	—	—	—	1	1
Chile (including the Juan Fernández Islands)	9	2	—	—	2	—	2
Venezuela	104	—	4 [2]	—	4 [2]	2	6
Brazil	90	—	1	3 [1]	4 [1]	4	8
Uruguay	6	—	—	—	—	1	1

Historical Extinctions

Only two hummingbird species are known with certainty to have become extinct in modern times, both in the Caribbean. This is equivalent to 0.5% of all hummingbird species, an extinction ratio that is considerably lower than the figure of 1.4% for all birds globally.

Brace's Emerald *Chlorostilbon bracei* The only specimen, a male, was collected about 3 miles from Nassau, on the island of New Providence, Bahamas, in July 1877 by the eponymous Lewis J. K. Brace. Doubts have been raised as to whether this is a good species, not least because other collectors (including near-contemporaries) failed to find it, but the subsequent discovery of fossil bones confirms that Brace's Emerald once ranged throughout New Providence and likely beyond. It appears to have survived in the dense vegetation that surrounded Nassau at the time of discovery, but to have become extinct shortly afterwards.

Caribbean Emerald *Chlorostilbon elegans* Even less is known about this species than about Brace's Emerald. It was described (as one of a batch of no fewer than 22 new hummingbirds, the skins of which had been languishing for various lengths of time in the British Museum) by John Gould in 1860 on the basis of a single specimen of unknown provenance, usually speculated to have been the Bahamas or Jamaica. Doubts about its validity have now been allayed, but there have been no further records and the species is almost certainly extinct.

Species on the brink

The majority of hummingbirds that are considered to have had stable or increasing populations over recent decades are those that are more tolerant of, or adaptable to, changes in their environment. As discussed in the chapter *Biogeography and Biodiversity: The Hummingbirds' Realm* (*page 115*), however, many hummingbirds are confined to very specific habitats or occur in just one small geographic area, and it is these species that are most likely to be adversely affected by habitat loss or degradation. It is therefore not surprising that all nine species of hummingbird considered to be at greatest risk of extinction—those categorized as Critically Endangered—have extremely small geographic ranges.

Although the number of hummingbirds currently considered to be under threat is undoubtedly concerning, just two species have been formally declared extinct, both last reported in the mid- to late-19th century (see *above*). Many of the hummingbirds that are currently considered to be of conservation concern could, however, face a similar fate unless urgent action is taken to safeguard their future. The remainder of this chapter provides an overview of the threatened hummingbirds that occur in each of the countries of the Americas, moving from north to south, together with a series of case studies covering principally those species that BirdLife International categorizes as Globally Threatened.

Black-backed Thornbill *Ramphomicron dorsale* (Endangered) | Colombia

USA and Canada

NT Rufous Hummingbird [winters in Mexico]

Several of the species of hummingbird that are known to have increasing populations occur in the USA and Canada: Black-chinned Hummingbird, for example, has increased by two-thirds since 1970, and Ruby-throated Hummingbird and Anna's Hummingbird have more than doubled in number over the same period—probably, as explained in *Biogeography and Biodiversity: The Hummingbirds' Realm*, as a result of the increased availability of nectar sources (see *page 176*). The **Rufous Hummingbird** is the only species in the region that is currently considered to be of conservation concern, as its population appears to be undergoing a moderately rapid decline. Even in such a well-studied part of North America, however, the reasons for this are not yet fully understood and more research is needed.

Rufous Hummingbird *Selasphorus rufus* | Canada

Caribbean Islands

NT Bee Hummingbird (see also *page 209*)

Although 17 hummingbirds are endemic to the Caribbean islands, none is considered Globally Threatened and just one species, the tiny **Bee Hummingbird**, which occurs only on Cuba, is of conservation concern. This species was once a widespread inhabitant of open woodland and thorn scrub on the island, but has become increasingly uncommon and localized as these habitats have been cleared for agriculture. It is now encountered primarily along remaining woodland edges, although it will occasionally also visit flowers and bird feeders in rural gardens.

Bee Hummingbird *Mellisuga helenae* | Cuba

Mexico

CR Short-crested Coquette
EN Oaxaca Hummingbird
VU Mexican Woodnymph
NT Tres Marias Hummingbird
NT White-tailed Hummingbird
NT Mexican Sheartail
NT Rufous Hummingbird [breeds USA/Canada]

Of the 60 hummingbird species recorded in Mexico, three are Globally Threatened. Of these, the one most at risk is the **Short-crested Coquette**, which is restricted to a tiny area of less than 50 km² (19 square miles) along a 25-km (16-mile) stretch of road in the Sierra de Atoyac, in the Sierra Madre del Sur of southwestern Mexico. Its semi-deciduous forest habitat was cleared rapidly during the 1990s, and the few remnant patches are unprotected and threatened by further deforestation for agriculture, mainly for corn and coffee cultivation. Recent attempts to assess this species' conservation status have been frustrated by the lack of access owing to the use of the area for narcotics cultivation, but the population is estimated to be fewer than 1,000 breeding birds.

The **Oaxaca Hummingbird** is restricted to a range of about 2,000 km² (770 square miles) in the Sierra de Miahuatlán, an isolated range of mountains at the eastern extremity of the Sierra Madre del Sur in the state of Oaxaca. It is a species primarily of high-altitude, humid forests, but is occasionally found at lower altitudes and may be a seasonal migrant. The breeding population is estimated at 600–1,700 birds, and from calculations of habitat loss it is thought that this is about half the number of a century ago.

The favored forest habitat of both Short-crested Coquette and Oaxaca Hummingbird is not represented within the Mexican Natural Protected Areas System and there is a pressing need to protect what little remains. Creation of a protected area in the Sierra de Atoyac is a particular priority if the future of the Short-crested Coquette is to be safeguarded.

Unlike the two previous species, the **Mexican Woodnymph** is found farther north in Mexico, in the southern canyons and foothills of the Sierra Madre Occidental, where it is fortunate to have some degree of protection in a national park and two Biosphere Reserves. Nevertheless, its population, estimated at fewer than 20,000 individuals, is thought likely to decline further because of the continuing loss of habitat. This hummingbird favors humid forest, especially along shady streams, and, although its ecological requirements are poorly known, it does seem to have been able to adapt to some degree to changes in its environment, and has even been found exploiting coffee plantations.

Short-crested Coquette *Lophornis brachylophus* | Mexico

Oaxaca Hummingbird *Eupherusa cyanophrys* | Mexico

Honduras

VU Honduran Emerald

Until its rediscovery in 1988 in the Aguán Valley, in northern Honduras, the **Honduran Emerald** was known from only 11 specimens, and was even feared extinct. It is therefore hardly surprising that it was assessed as Critically Endangered until the discovery of further populations in 2007 and 2008, as a result of which it was reassessed as Endangered. Yet more locations for this species have subsequently been found and the population is now estimated to be in the range 10,000–20,000 individuals: as a result, at the end of 2020 it was recategorized as Vulnerable. The species' stronghold is still the upper Aguán Valley, which contains the largest extent of deciduous thorn forest in Honduras (85 km^2/33 square miles), within which the bird is relatively common, although it is found also in other thorn forests and semi-deciduous woodland in the interior valleys of Honduras.

Although the discovery of a relatively large population is encouraging, this hummingbird is still imperiled by the threat shared by most hummingbirds: habitat destruction. The thorn forests across much of Honduras have been cleared with bulldozers for agriculture (especially for pineapples, but also bananas and rice) and pasture; this mechanized clearance causes permanent damage to the habitat, which could otherwise recover if cleared by hand. Fortunately, the Honduran Emerald seems able to tolerate a degree of disturbance, such as through grazing and small-scale agriculture, provided that fragments of its preferred habitat survive. In 2005, the Honduran government designated 12 km^2 (4.6 square miles) as a *Honduran Emerald Species and Habitat Management Area* west of the town of Olanchito, about half of which

Honduran Emerald *Amazilia luciae* | Honduras

is prime dry-forest habitat. The reserve is popular with bird tourists keen to visit the only easily accessible site for the species, the revenue generated contributing to its conservation.

Costa Rica

CR Guanacaste Hummingbird

EN Mangrove Hummingbird

The curious case of the **Guanacaste Hummingbird** was described in 1989 by the celebrated Costa Rica field-guide authors Stiles & Skutch as the "foremost ornithological mystery of Costa Rica." This hummingbird is known from only one specimen that was collected in 1895 on the south slopes of the Miravalles Volcano in northwest Costa Rica and deposited in the Natural History Museum at Tring, UK. Notwithstanding confirmation of its species status by eminent ornithologists of the time, 15 years later it was suggested that it might better be considered a geographically isolated subspecies of Indigo-capped Hummingbird (see *page 269*), a Colombian species not found within 1,000 km (610 miles) of the volcano; the latter view prevailed, resulting in its subsequent 'lumping' with that species. At the turn of the 21st century, however, a re-examination of the specimen concluded that it was indeed a valid species. Despite this, some taxonomic authorities take the view that it is a hybrid.

Since 2016, Guanacaste Hummingbird has been regarded as a species by BirdLife International and assessed as Critically Endangered, only the fact that systematic searches had not yet been conducted saving it from being formally categorized as either Possibly Extinct or even Extinct. In 2020 and 2021, naturalist Ernesto Carman carried out targeted searches in the area where the specimen was collected on Miravalles Volcano in search of this mysterious bird. Large numbers of both Purple-throated Mountain-gem and the blue-vented form of Steely-vented Hummingbird—two possible parents of a putative hybrid—were found, often interacting at the same patches of flowers, but nothing resembling Guanacaste Hummingbird was seen. A large extent of habitat, however, remains to be surveyed. Whatever the outcome, its habitat has received formal protection through the creation in 2019 of the Miravalles Jorge Manuel Dengo National Park, giving hope that, if any birds do remain, their population might be safeguarded.

Mangrove Hummingbird *Amazilia boucardi* | Costa Rica

Found only on the Pacific coast of Costa Rica, the **Mangrove Hummingbird** is one of the most range-restricted hummingbird species in Central America. As its name suggests, it is confined to the coastal zone, where it occurs chiefly in mangroves, although it will visit adjacent habitats. Unfortunately, mangrove forests are continually being lost to development pressures, particularly to create salt and shrimp ponds, and may also be adversely affected by sea-level rise exacerbated by climate change.

Panama

EN Glow-throated Hummingbird

NT Pirre Hummingbird [also Colombia]

NT Violet-capped Hummingbird [also Colombia]

The **Pirre Hummingbird** is an uncommon and local resident of mountain foothills in the Darién region of Panama and extreme northern Colombia, with most recent sightings from two areas of Cerro Pirre in Panama. Its small range is remote and mountainous and therefore spared much human intervention, but the threat of completion of the Panamerican Highway to connect Central and South America has long hung over this extraordinary wilderness.

Pirre Hummingbird *Goethalsia bella* | Panama

The **Glow-throated Hummingbird** is known only from two mountain regions, the Serranía de Tabasará and the highlands of the Azuero Peninsula, in western and central Panama, respectively. It occurs mainly at altitudes of over 1,200 m (3,900 ft) in the lower levels of humid broadleaf forest. Much of its habitat is inaccessible, but even with concerted efforts to track down the bird there are only a few dozen modern-day sightings, and little is known about its ecology. Its population has been crudely estimated to be 2,200–11,000 breeding birds, and declining. In the historical core of this species' range on Cerro Santiago, deforestation for cattle pasture continues. Although Glow-throated Hummingbird is apparently able to survive in secondary and degraded forest, it cannot tolerate replacement of the understory with crops such as coffee.

Venezuela

EN **Venezuelan Sylph**
EN **Perija Metaltail** [also Colombia]
EN **Perija Starfrontlet** [also Colombia]
EN **Scissor-tailed Hummingbird**
NT **Coppery-bellied Puffleg** [also Colombia]
NT **White-tailed Sabrewing** [also Tobago]

Venezuela's hummingbirds of conservation concern are found in two geographic areas: the Eastern Coastal Cordilleras and the Andes. The Paria Peninsula, in the far northeast of the country, is home to the extraordinary **Scissor-tailed Hummingbird**, a *Heliconia*- and ginger-feeding denizen of the unique cloud forests of the area, with a population numbering 5,000–6,000 individuals. Although much of the range of this species has been covered by the 375 km^2 (145 square miles) of Paria Peninsula National Park since 1978, appropriate management and legal enforcement are minimal.

The **Venezuelan Sylph** is found farther west along the Eastern Coastal Cordilleras, where it is restricted to the humid, wet forests and coffee plantations of the Caripe Highlands. There has, however, been extensive deforestation across its extremely small range and the population is currently estimated to be between just 1,500 and 7,000 mature individuals.

The remote Andean mountains of the Sierra de Perijá, which straddle part of the northern border between Venezuela and Colombia, are home to both the **Perija Metaltail** and **Perija Starfrontlet**, two birds that were virtually unknown before the turn of the millennium and whose basic biology is still largely a mystery. The metaltail is restricted to two areas of high-altitude *páramo* and elfin forest above 2,400 m (7,900 ft); the starfrontlet occurs in taller cloud-forest habitat within the narrow altitudinal range of 2,550–3,100 m (8,400–10,200 ft) and is estimated to number fewer than a thousand mature individuals. The habitats in which these hummingbirds are found are under severe pressure from subsistence and commercial agriculture, as well as narcotics cultivation, and as a result both species are presumed to be declining.

Scissor-tailed Hummingbird *Hylonympha macrocerca* | Venezuela

Colombia

CR **Blue-bearded Helmetcrest**
CR **Gorgeted Puffleg**
CR **Santa Marta Sabrewing**
EN **Black-backed Thornbill** (see *page 194*)
EN **Perija Metaltail** [also Venezuela]
EN **Colorful Puffleg**
EN **Glittering Starfrontlet**
EN **Perija Starfrontlet** [also Venezuela]
EN **Sapphire-bellied Hummingbird**
VU **Ecuadorian Piedtail** [also Ecuador and Peru]
VU **Buffy Helmetcrest**
VU **Black Inca**
VU **Pink-throated Brilliant** [also Ecuador]
VU **Santa Marta Blossomcrown**
VU **Tolima Blossomcrown**
NT **Wire-crested Thorntail** [also Ecuador and Peru]
NT **Hoary Puffleg** [also Ecuador]
NT **Black-thighed Puffleg** [also Ecuador]
NT **Coppery-bellied Puffleg** [also Venezuela]
NT **White-tailed Starfrontlet**
NT **Chestnut-bellied Hummingbird**
NT **Pirre Hummingbird** [also Panama]
NT **Violet-capped Hummingbird** [also Panama]

Several Globally Threatened species of hummingbird have largely eluded detection in recent years, perhaps the most enigmatic of these being the distinctive **Gorgeted Puffleg**, the only member of the genus in which the male shows a brilliant bicolored gorget. This species was first encountered in 2005, in the remote Andean mountain range of the Serranía del Pinche of southwestern Colombia, and was described as a new species in 2007. It appears, however, not to have been seen subsequently. It was found in wet cloud forest and stunted elfin forest at 2,600–3,200 m (8,500–10,500 ft) and its global range is probably less than 10 km² (3.9 square miles) in extent: based on the known population densities of other pufflegs, it has been estimated that only around 600 mature individuals remain. The main threat is the inexorable loss of habitat driven by the expansion of subsistence agriculture and narcotics cultivation. Fortunately, the area where this puffleg occurs falls within the Serranía del Pinche Protective Forest Reserve (which was designated in 2008), and in 2011 the Colombian non-profit environmental organization Fundación ProAves established an area of 7.5 km² (2.9 square miles) as the Gorgeted Puffleg Nature Reserve, dedicated primarily to the conservation of this hummingbird. Nevertheless, the species remains very poorly known, and the most urgent conservation challenge is to determine its precise distribution and population size and to develop a species-specific management plan.

Sapphire-bellied Hummingbird *Amazilia lilliae* | Colombia

The **Sapphire-bellied Hummingbird** is restricted to a stretch of around 100 km² (39 square miles) along the Caribbean coast of Colombia, an area which has been subject to a range of significant development pressures over recent decades. This hummingbird particularly favors coastal mangroves, but is also known to make local seasonal movements, occasionally occurring in xerophytic scrub and sometimes showing a preference for forests of Coral Bean Tree *Erythrina fusca* while they are in flower, possibly acting as a pollinator. The total number of mature individuals is estimated to be fewer than 250, and even this tiny population is suspected still to be declining rapidly owing primarily to continuing habitat loss.

Despite targeted searches, there were no records of **Blue-bearded Helmetcrest** for almost 70 years until, in 2015, three individuals were observed in a remote part of the Sierra Nevada de Santa Marta, in northern Colombia. As so little is known about this species, it has not been possible to calculate its population size or trend, but fewer than 250 mature individuals are assumed to survive. The *páramos* in which this hummingbird lives are subject to regular burning to create fresh pasture for livestock, and as a result the plant *Libanothamnus occultus* (Asteraceae), the flowers of which are suspected to be its principal food source, is also severely threatened.

The **Santa Marta Sabrewing** likewise is restricted to the Sierra Nevada de Santa Marta, where historically it was recorded at lower elevations than the Blue-bearded Helmetcrest. Over the last 75 years, however, this hummingbird has been recorded with certainty only once, in 2010, and almost nothing is known of its biology.

For more than 50 years the **Glittering Starfrontlet** was known only from a single specimen collected in 1951. After its rediscovery in 2004, this hummingbird was immediately assessed as Critically Endangered on the assumption that fewer than 250 breeding birds remained. Since then, however, research suggests that its population may be as much as ten times larger and, as a consequence, the species was recategorized as Endangered in 2020. It occurs in 270 km² (104 square miles) of transitional habitat associated with *páramo*-elfin forest in the western Andes of Colombia, where it is under severe threat of habitat loss owing to continuing deforestation for farming and the potential expansion of mining operations. Its habitat is likely to be highly susceptible also to the impact of climate change, including an increased frequency of droughts and subsequent forest fires.

The subtly beautiful **Black Inca** is found only in Colombia, where it inhabits principally mature, humid montane forest at an altitude of 1,000–2,800 m (3,300–9,200 ft) on the slopes of the eastern cordillera of the Andes not far from Bogotá. Although its range has recently been found to be more extensive than previously thought, covering around 17,100 km² (6,600 square miles), its distribution is highly fragmented, and suitable patches of habitat are decreasing in size and quality through ongoing degradation and clearance for agriculture. The total number of mature individuals is currently estimated to be fewer than 9,000, and decreasing.

There is a 'better-news' story associated with the **Chestnut-bellied Hummingbird**, which is yet another species endemic to Colombia. Restricted to the drier parts of the Magdalena Valley in the department of Santander, it was for many years believed to have a tiny population and therefore to be highly threatened—and as a result was categorized as Endangered and even Critically Endangered for many years. It has recently been found, however, to occur over a more extensive area than was previously assumed and to have a population of around 16,000 mature individuals. Although this figure may actually be a considerable underestimate, there is evidence that the population is declining and the species is therefore considered to be Near Threatened.

Glittering Starfrontlet *Coeligena orina* | Colombia

Black Inca *Coeligena prunellei* | Colombia

Chestnut-bellied Hummingbird *Amazilia castaneiventris* | Colombia

Ecuador

CR **Blue-throated Hillstar** (see *page 207*)
CR **Turquoise-throated Puffleg**
EN **Royal Sunangel** [also Peru]
EN **Violet-throated Metaltail**
EN **Black-breasted Puffleg**
VU **Ecuadorian Piedtail** [also Colombia and Peru]
VU **Pink-throated Brilliant** [also Colombia]
VU **Esmeraldas Woodstar**
NT **Wire-crested Thorntail** [also Colombia and Peru]
NT **Hoary Puffleg** [also Colombia]
NT **Black-thighed Puffleg** [also Colombia]
NT **Napo Sabrewing** [also Peru]
NT **Little Woodstar** [also Peru]

Violet-throated Metaltail *Metallura baroni* ♀ | Ecuador

Like the Gorgeted Puffleg of Colombia, the **Turquoise-throated Puffleg** is another enigmatic and very little-known hummingbird that may still occur in a tiny area of Ecuador. The evidence for its existence, however, amounts to just four specimens from the northwest of the country and two trade skins labeled 'Bogotá' and therefore possibly obtained in southwest Colombia: only one of the specimens, collected in 1850, carries locality information. The species has not been recorded in the intervening 170 years, although there was an unconfirmed sighting from Chillo Valley near Quito in 1976. Targeted searches at several sites in 1980 and 2004–2005 have failed to find any trace of the bird, but, as there is a chance, albeit slim, that it may still exist, it has not been declared extinct.

Thankfully, the **Black-breasted Puffleg** is rather better known. It is restricted to the high Andes of northwest Ecuador, and although there are numerous specimens of this hummingbird in museum collections, suggesting that it may once have been fairly common, it is now extremely rare, with a population of 100–150 birds. Its global distribution is estimated at 68 km² (26 square miles), split into two subpopulations: one on the northwestern flanks of two adjacent volcanoes, Pichincha and Atacazo, in Pichincha province; and another (rediscovered in 2006) in the Toisán Range, Imbabura province. Its presence on Atacazo Volcano is, however, based only on three males collected from the locality in 1898; there have been no subsequent sightings, despite recent surveys, and there are fears that it no longer occurs here.

Black-breasted Puffleg *Eriocnemis nigrivestis* | Ecuador

Traditionally, the Black-breasted Puffleg was thought to have been confined to the crests of mountain ridges dominated by elfin forest with abundant epiphytes and ericaceous plants, but it is now known to visit a broad range of flowering plants, and has recently been found in shrubby forest borders, on steep slopes with stunted vegetation, and even in the interior of mature montane forest. Like many montane hummingbirds, it

Esmeraldas Woodstar *Chaetocercus berlepschi* | Ecuador

is an altitudinal migrant, breeding at 2,700–3,500 m (8,900–11,500 ft) during the November–February rainy season, when flowers bloom at high elevation, and dropping down to as low as 1,700 m (5,600 ft) during the dry season. More than 90% of the potential habitat within its probable historic range has been degraded or destroyed by charcoal-burning, conversion to cattle pasture and potato cultivation, and only a few patches of forest remain. Fortunately, charcoal production on the slopes of Pichincha Volcano is now prohibited, and access to the area is controlled.

In 2001, the Ecuadorian non-governmental organization Fundación Jocotoco acquired almost 950 hectares (2,350 acres) of prime Black-breasted Puffleg habitat on the flanks of Pichincha Volcano and established the Yanacocha Reserve. Subsequently, there has been considerable local interest, with significant press and television coverage, particularly since the reserve is easily accessible on a day trip from Quito. The reserve has slowly expanded and now comprises 12 km^2 (4.6 square miles) of habitat, in part thanks to funding from a carbon offsetting scheme, some forward-thinking companies having used this initiative to counterbalance the carbon emissions of their entire business. Protection of these remnants of high-altitude forest also benefits the citizens of parts of nearby Quito by helping to ensure a continuing source of fresh water.

The **Violet-throated Metaltail** has a very small range at an altitude of 3,100–4,000 m (10,200–13,100 ft) in the Andes of south-central Ecuador, within which it is known from just a few locations. It occurs in *Polylepis* woodland, shrubby *páramo* and the upper edge of montane forest—habitats that are continually being lost, fragmented or degraded as a result of human activities. The population of this hummingbird is estimated to be in the region of 600–1,700, and is continuing to decline.

During its breeding season, from October to March, the tiny **Esmeraldas Woodstar** inhabits a very small and highly fragmented area of semi-deciduous to moist evergreen lowland forest along the coast of Ecuador from Esmeraldas in the north to Guayas in the south. It is not known for certain where it spends the rest of the year, although it is believed to move to higher altitudes. Its population is currently estimated to be fewer than 2,700 mature individuals, and is declining at a rate of around 25% every ten years as a result of continuing habitat loss and degradation.

Wire-crested Thorntail is sparsely distributed and generally uncommon in humid forest between 400 m and 1,500 m (1,300–4,900 ft) along the edge of the Amazon Basin in Colombia, Ecuador and Peru (and there have also been recent records from northwest Bolivia). It does not seem able to tolerate secondary habitats and is therefore highly susceptible to deforestation. For this reason, it was reclassified as Near Threatened in 2012 and, sadly, may be reassessed as Globally Threatened in the near future.

Ecuadorian Piedtail *Phlogophilus hemileucurus* | Ecuador

Wire-crested Thorntail *Discosura popelairii* | Peru

Peru

EN **Royal Sunangel** [also Ecuador]
EN **Grey-bellied Comet**
EN **Marvelous Spatuletail** (see *page 208*)
VU **Ecuadorian Piedtail** [also Colombia and Ecuador]
VU **Purple-backed Sunbeam**
NT **Koepcke's Hermit**
NT **Wire-crested Thorntail** [also Colombia and Ecuador]
NT **Peruvian Piedtail**
NT **White-tufted Sunbeam**
NT **Napo Sabrewing** [also Ecuador]
NT **Little Woodstar** [also Ecuador]

The **Royal Sunangel** is known to occur in just a few scattered localities in northern Peru and extreme southeast Ecuador. It is most numerous in so-called 'elfin scrub', a dry, stunted vegetation of poor, sandy soils on mountain ridges. Although these soils lack the nutrient levels to support agriculture or cattle-ranching, much of the sunangel's habitat is now surrounded by cultivated land, and the expansion of commercial mining operations and road-building threaten the little that remains. At present, the population is estimated to be fewer than 10,000 mature individuals.

The **Grey-bellied Comet** is found exclusively in a small number of sites in northern Peru, where it is inexplicably rare. Estimates put the total population at fewer than 1,000 breeding birds, and declining. Its favored habitat is semi-arid montane scrub above 2,750 m (9,000ft), but areas in much of its known range are burnt frequently for agriculture or to create fresh pasture for livestock. The species is thought to have always been tolerant of some degree of habitat modification by humans, but whether it is able to complete its life-cycle or occur at normal densities in disturbed areas is not yet known.

Despite its name, the **Ecuadorian Piedtail** ranges from Colombia to Peru. It inhabits humid forest between 600 m and 1,500 m (2,000–4,900 ft) along the eastern foothills of the Andes and is declining as a result of deforestation. It was reclassified as Vulnerable in 2012 for this reason.

Royal Sunangel *Heliangelus regalis* | Peru

Grey-bellied Comet *Taphrolesbia griseiventris* | Peru

Chile

CR Chilean Woodstar [also Peru but no recent records]

CR Juan Fernandez Firecrown

Chilean Woodstar is confined to a few oasis river valleys in the Andean foothill deserts of northern Chile. Despite being very common in the early 20th century and still abundant in the 1980s, by 2017 the population had plummeted to around just 210 adult birds. This hummingbird is thought to require contiguous areas of suitable habitat along the narrow valley floors, enabling it to make local altitudinal movements, and it probably has a preference for native flowers rather than ornamental garden plants and crops. Most of the area in which it previously occurred, however, has been irrigated and converted to intensive agriculture, and this, combined with the impact of the heavy use of pesticides to control crop pests, and possibly competition with another hummingbird, the Peruvian Sheartail (which has colonized since 1971), may be the reason for this species' demise. The good news is that there are now conservation initiatives underway to raise awareness among local communities of the plight of this hummingbird and to protect and restore suitable habitat.

The **Juan Fernandez Firecrown** is endemic to Isla Robinson Crusoe, one of the Juan Fernández Islands, located in the Pacific Ocean 670 km (416 miles) from the mainland coast of Chile, where it is confined to about 11 km^2 (4.2 square miles) of remaining habitat. The most recent survey, in 2011, estimated the population to be of 740 birds, and calculated that it was decreasing by 1–9% per decade. The subspecies *leyboldi*, which was endemic to Isla Alejandro Selkirk, another island in the archipelago, is already extinct, having last been recorded in 1908.

Much research has been carried out into the biology of the Juan Fernandez Firecrown in order to understand the factors affecting its conservation. It is believed to have coevolved with the native flora, of which two-thirds of the constituent species are endemic; in late summer, for example, it feeds preferentially on the endemic Cabbage Tree *Dendroseris litoralis*. Unfortunately, 90% of the native vegetation on Isla Robinson Crusoe has been lost. The extent of exotic *Acacia*, *Eucalyptus*, cypress and pines is increasing at the expense of native habitat, while the remainder has been invaded by introduced plants such as Maqui/Chilean Wineberry *Aristotelia chilensis*, Elm-leaf Blackberry *Rubus ulmifolius* and Chilean Guava *Ugni molinae*. In addition, introduced grazing mammals, principally feral goats, cattle and rabbits, have had an impact by modifying the natural vegetation, while coatis, cats, mice and rats are potential predators, and Austral Thrushes, also introduced, have been observed preying on nests. The Green-backed

♂ (*top*) and ♀ (*bottom*) **Juan Fernandez Firecrown** *Sephanoides fernandensis* look so different, both in size and in coloration, that they were originally described as two separate species. | Juan Fernández Islands, Chile

Chilean Woodstar *Eulidia yarrellii* | Chile

Firecrown, which is common on mainland Chile, colonized Isla Robinson Crusoe relatively recently, probably over the last few centuries, and being the Juan Fernandez Firecrown's closest relative it exhibits similar morphology and ecological requirements. This species now competes with the Juan Fernandez Firecrown to the extent that each species excludes the other, the newcomer currently outnumbering the endemic by four to one. There are ongoing efforts through a Chilean government program to remove alien invasive plants and mammalian predators, and Cabbage Trees have been planted in and around settlements.

Bolivia

NT Dot-eared Coquette
[one unconfirmed record of this Brazilian endemic]

DD Coppery Thorntail

The **Coppery Thorntail** is the least known of all hummingbirds, all knowledge being derived from two specimens of vague provenance, usually presumed 'Bolivia', collected before 1852. As there is still a possibility, albeit slim, that this species has been overlooked in the wild, it is categorized as Data Deficient rather than Extinct.

The **Wedge-tailed Hillstar** was previously categorized as Near Threatened but was downlisted to Least Concern in the 2021 Red List update, after its potential global population was recalculated using densities derived from studies of White-sided Hillstar and estimated to be 10,000–30,000 breeding birds. It is a rather localized and scarce inhabitant of dry *Polylepis* woodland and scrub on the upper slopes of the eastern Andes, principally of Bolivia, although it occurs also in the extreme northwest of neighboring Argentina. Ongoing loss of its favored habitat was formerly considered to be a significant threat, but the species seems able to survive in areas severely degraded by burning and overgrazing, provided that some patches of bushy native vegetation remain.

Wedge-tailed Hillstar *Oreotrochilus adela* | Argentina

Brazil

EN Long-tailed Woodnymph
VU Hook-billed Hermit
VU Tapajos Hermit
VU Dry-forest Sabrewing (see *page 239*)
NT Hooded Visorbearer
NT Dot-eared Coquette
[perhaps also Bolivia, where status uncertain]
NT Festive Coquette
NT Diamantina Sabrewing

The **Hook-billed Hermit** was once found throughout the Atlantic Forest of Brazil, from the State of Bahia in the north to the State of Rio de Janeiro in the south, a distance of more than 1,000 km (620 miles). Like many hermits, it prefers to remain inside humid forests, favoring particularly streambeds with

Hook-billed Hermit *Glaucis dohrnii* | Brazil

flowering *Heliconia*. There has, however, been extensive deforestation throughout the historic range of this hummingbird, and it is now restricted to a few tiny and fragmented subpopulations, with the total number of mature individuals currently estimated to be fewer than 1,000, and probably still declining owing to continuing habitat loss.

The **Long-tailed Woodnymph** is confined to the extreme northeast of Brazil, where it is restricted to lowland habitats, primarily coastal rainforest and *cerrado*, but it may occur also in plantations and parks. The population, which is estimated to number no more than 2,500 mature individuals, has become extremely fragmented as a result of severe loss of habitat, with probably only around 250 birds in each subpopulation. Although it occurs in a few protected sites, very little is known about its ecology or its ability to persist in degraded and isolated patches of habitat.

The **Tapajos Hermit**, which was recognized as a species only in 2009, is restricted to undisturbed primary forest in Amazonian Brazil between the Tapajós and Xingu rivers. Unfortunately, the area in which it occurs has one of the highest rates of deforestation in the Amazon—a situation that is highly likely to be further exacerbated by the paving of a trans-Amazonian highway between the southern city of Cuiabá, in the state of Mato Grosso, and the city of Santarém, in the Amazonian state of Pará, opening the area up for even more settlement. No information is available on the population size of this hummingbird, but it is presumed to be declining owing to the severity of ongoing habitat loss.

Despite occurring across a large swathe of the Amazon Basin, the **Dot-eared Coquette** is a very poorly known hummingbird. Its habitats include forest edge, savanna and *cerrado* (dry savanna woodland) up to 500 m (1,600 ft) in altitude, but accurate information on its population size is lacking. In light of the projected rate of continuing deforestation across the Amazon Basin, primarily as land is cleared for cattle-ranching and soy production, facilitated by expansion of the road network, the species is, however, considered to be Near Threatened, having been downlisted from Vulnerable in the 2021 Red List update.

Long-tailed Woodnymph *Thalurania watertonii* | Brazil

Unexpected discoveries

Although our knowledge of hummingbirds is continually increasing, this due in no small measure to the exponential growth in the number of professional and keen amateur ornithologists, there seems little doubt that plenty of surprises still await us. For example, as recently as 2017, a well-known Ecuadorian ornithologist, Dr Francisco Sornoza-Molina, made a brief visit to the high Andean *páramos* of Cerro de Arcos in El Oro province, southwest Ecuador, an area poorly explored historically by ornithologists. During this visit he had the good fortune to observe and photograph an immature male hummingbird that looked similar to Green-headed Hillstar and Ecuadorian Hillstar (*page 218*) but did not fit precisely any of the six known species in that genus (*Oreotrochilus*). Subsequent visits by a larger team revealed that both the males and the females of this 'mystery' hummingbird possessed unique combinations of plumage characters, suggesting that the hillstar was a new species, and it was formally described in the following year as Blue-throated Hillstar. Analyses of mitochondrial DNA have indicated that the new species, contrary to appearances, is most closely related genetically to Green-headed Hillstar and Black-breasted Hillstar (*below right*), rather than to Ecuadorian Hillstar.

Further visits to areas of potentially suitable habitat for the Blue-throated Hillstar established that it was present only in the Chilla-Tioloma-Fierro Urcu Cordillera. Since this hummingbird was apparently confined to a single mountain range, had a global distribution of less than 100 km^2 (38 square miles) and a population estimated to be in the low hundreds, and its habitat was under threat, it was evaluated as Critically Endangered by the researchers who discovered it. The species was subsequently recognized by BirdLife International and was categorized as Critically Endangered in the 2021 Red List update.

The main threat faced by this hummingbird is the burning, grazing and conversion of the *páramos* to crops and pastureland, but, more worryingly perhaps, a large proportion of its known range is also threatened by gold-mining concessions. The unexpected discovery of a new species of hummingbird, however, generated so much excitement that a number of national and international conservation NGOs were able very quickly to raise sufficient funds to purchase land in the area where the birds had been observed. In addition, there are local initiatives underway that aim to designate much of the hillstar's habitat as a water-supply protection area.

Blue-throated Hillstar *Oreotrochilus cyanolaemus* | Ecuador

Black-breasted Hillstar *Oreotrochilus melanogaster* | Peru

Marvelous Spatuletail *Loddigesia mirabilis* (Endangered) | Peru

Looking to the future

As we have seen, the main threat to hummingbirds is the disappearance and deterioration of the wild places in which they live. These birds and their habitats have evolved together over millions of years, yet within an evolutionary blink of an eye the threat of extinction is becoming all too real. Their persecution has thankfully been consigned to history, and nowadays most people who come into contact with hummingbirds value them for their intrinsic beauty, as a source of inspiration, and as something that enriches our lives. Nobody wants to see them go. It is simply the conflict between the habitats upon which they depend and our own demands for more space, more food, more farmland, more raw materials and more land for development.

Two hummingbird species have become extinct in recent times, and others now face that fate: from a human perspective, it is we who face their irreplaceable loss. But surely it is within our power to conserve those that remain, to give them enough space to live out their busy lives alongside our own. We know that it is possible to reverse recent trends towards extinction, and indeed with the most threatened of the world's hummingbirds in particular there are people working tirelessly to ensure that they have a future. We should not, however, simply rely on these dedicated few, and we can all play our part, even if indirectly, by taking more care of our planet, acknowledging the difficult challenges of climate change and reducing our demands on natural resources and energy. There is also the opportunity to help to increase our knowledge of hummingbirds, even if simply by reporting sightings to local conservation organizations or online via, for example, eBird (see *page 277*), which will give conservationists a better chance of coming up with the answers needed to solve some seemingly intractable problems. Hummingbirds deserve the best we can give them, and in return they will show us the best that the natural world has to offer. If we can start to take meaningful steps towards such conservation, all of us can then be optimistic that these beautiful little flying jewels will still stir our emotions in the real world and not just as a series of stunning pictures in a book.

FACING PAGE **Bee Hummingbird** *Mellisuga helenae* (Near Threatened) | Cuba

The future of these incredible birds is in our hands

Hummingbirds, and the habitats upon which they depend, face real threats in the years to come. It is not too late, however, to save these amazing birds. We can all help by making direct contributions to conservation organizations, by supporting conservation indirectly through well-run ecotourism initiatives, or simply by reducing our collective impact upon this planet. Anything we can do will be worthwhile, as a world without these magical little flying jewels would be a very sad place indeed.

In Pursuit of Hummingbirds: A Photographic Journey

Glenn Bartley

The Beginning of an Obsession

I have been fascinated by hummingbirds for as long as I can remember. Where I grew up in Canada, we had only a single species of hummingbird, the Ruby-throated Hummingbird, which would from time to time grace us with its presence. For me as a young person this was always an exciting occurrence. Seemingly from out of nowhere our backyard friend would descend into our world, zipping from flower to flower with the utmost determination. It is an amazing thing to observe a hummingbird up close. The speed, grace and accuracy with which these birds move never cease to amaze me. They are truly captivating to watch and seem perpetually to tease their onlookers with fancy acrobatic maneuvers and flashes of brilliant iridescence. Alas, it is often the case that these birds never allow us really to soak up their beauty and study their intricate details. Most of the time they disappear far too quickly, leaving their fans wanting more and hoping that they will return again soon.

Each summer when I was growing, up my father would take my brother and me camping at a lake in eastern Ontario. Several of the people in the campground set up sugar-water feeders for their tiny feathered companions. This provided me with my first chance really to watch and study these magnificent little birds. I can remember lying directly under one of the feeders and staring up at the sky waiting for a customer to arrive. When the hummingbirds finally came for a snack I could appreciate their subtle movements and amazing aerial capabilities. Still, though, I was left wanting more. The birds simply moved too fast to satisfy my curiosity and I began wondering if I could capture a photograph and 'freeze' the fleeting moments that they were offering me.

We had an extremely basic 35 mm film camera with us while we were camping. The equipment was completely inadequate to take any kind of decent image of such a difficult subject, but such a technical matter is not the kind of thing that a child, fascinated with nature, dwells upon. So, one summer's day I began to try to capture hummingbirds on film. Little did I know at the time that lying in ambush under the feeder and waiting for a target to appear would be the start of my quest to photograph the world's hummingbirds.

Ruby-throated Hummingbird *Archilochus colubris* | Canada

Many years have gone by since I first lay under that hummingbird feeder, and in that time I have spent thousands of hours in the field and taken tens of thousands of images of nearly 200 species of hummingbird. I am still just as amazed by these stunning little birds as I was all those years ago. If anything, the more I learn about hummingbirds and the more time I spend with them, the more I respect and appreciate them.

Dreams coming true: my first visit to the tropics

It would be many years before I would pick up a camera and try to photograph a hummingbird again. In fact it was not until I had completed my undergraduate university degree in my mid-twenties that I would get that chance. After all that studying I needed a break and wanted to pursue my growing interest in bird photography more

FACING PAGE **Anna's Hummingbird** *Calypte anna* | Canada

fully. Following graduation, I spent the first part of the year saving up money, paying off the school debt and planning for what would be the first of many trips to the 'New World' tropics. My destination of choice was the wonderful Central American nation of Costa Rica.

Costa Rica has a lot going for it from a birding and ecotourism standpoint. While not so diverse as some of the tropical Andean nations, it is astoundingly rich in terms of biodiversity for such a small country. For example, there are approximately 50 species of hummingbird in Costa Rica to chase after: plenty to keep a first-time tropical-bird photographer like me busy for six months! The country is also safe and easy to travel around by public transportation, which made it an ideal choice for me at the time.

I can still remember the very first photograph I took of a tropical hummingbird. Having arrived in Costa Rica's capital city of San José the night before, I set off by bus one morning for a biological research station in the heart of the lowland rainforest. It was not long before I had an impossibly beautiful purple and green male Crowned Woodnymph perched on a very complementary fuchsia-colored *Heliconia* flower. What a colorful introduction to tropical-bird photography!

Those who have studied the plates of field guides will be aware that, as good as the illustrations are, they are incapable of portraying just how stunning these birds are in real life. Part of what makes hummingbirds so magical is that their appearance is constantly changing. As they move their head from side to side and adjust their position, they gift onlookers with glittering flashes of iridescent splendor. There is something absolutely mesmerizing about the show that they put on for those fortunate enough to see them face to face. While our brain does allow us to realize the beauty of what we are seeing, however, it cannot quite soak up the details inherent in the ever-changing subjects sufficiently accurately to enable us truly to appreciate them. This is where a good photograph can be really magical.

If I am being totally honest, most of the images that I took on that first trip to Costa Rica were not very good or magical in any way. There were, however, a few standout shots that I am still proud of. In reality, though, that trip was a glimpse into the fascinating, complex and challenging world of tropical-bird photography. It was the trip that got me hooked on tropical birds, and I headed home humbled and determined to improve before I would return.

Crowned Woodnymph *Thalurania colombica* | Costa Rica

Learning how to photograph hummingbirds: studying behavior and developing new techniques

I learned a lot on my first trip to the tropics. Perhaps the most valuable lesson was the simple fact that I did not have the skill or equipment to photograph hummingbirds in the way I really wanted to. Nevertheless, all that time spent in the field in Costa Rica was certainly not wasted, as it allowed me to get a much better understanding of how hummingbirds behave day-to-day. I began to notice that some individuals were extremely habitual in the places where they chose to rest. The birds would return regularly to the exact same spot on a branch over and over again if they were feeding in an area or protecting a food source. Recognizing this predictable behavior is extremely beneficial when you are trying to photograph hummingbirds. I learned that, instead of hopelessly chasing them around, the best strategy is often simply to sit and wait. I also learned that sometimes it pays to set up some new perches close to a food source (ideally a sugar-water feeder). Once the birds get used to the new resting spots they can be extremely cooperative and, with a little planning, much more photogenic.

If only all hummingbirds were so cooperative to return repeatedly to the same spot! While this strategy works for some aggressive species, there are others that use different strategies altogether. One such strategy, employed largely by the hermits, is to feed along a so-called trapline. In other words, the birds have a mental map of a large number of flowering plants throughout the forest and will visit each of them in sequence a number of times throughout the day. The patient photographer can take advantage of this behavior. Photographing hummingbirds under the dark canopy of a tropical rainforest, however, almost always requires the use of artificial lighting.

On my first trip to Costa Rica I had only one flash for my camera, and the reality is that I barely knew how to use it. As a result, I relied almost exclusively on natural light for my images. When the conditions are ideal (the light is bright and at a low angle) this can work out perfectly well; in fact, this is my preferred way of photographing hummingbirds. The problem is that photographers frequently run into challenging, or impossible, lighting conditions. In the tropics, the sun rises quickly and almost immediately becomes very harsh—often, as soon as the sun clears the surrounding

Blue-throated Starfrontlet *Coeligena helianthea* | Colombia

mountains, the light is impossible to work with. Moreover, because hummingbirds are typically extremely active and dynamic subjects, it is always very challenging to keep a constant angle of light. Furthermore, many species prefer to stay in the dark parts of the forest or are found where cloudy weather is the norm. As a result, there simply is not sufficient natural light to get a fast enough shutter speed to photograph these speedy creatures. You can certainly capture successful images of perched hummingbirds in these conditions, and these are a great addition to a portfolio, but images of perched hummingbirds alone are not enough to represent these acrobatic little birds adequately. I knew that, in order to capture images of hummingbirds in flight that fully celebrate their beauty, I would need to come up with a better system.

Photography is at its core all about light. We use the light to 'paint' a subject in the real world into an image. The amount of light falling on a camera sensor or film depends on the size of the aperture behind the lens and the speed of the shutter that allows it to pass through in a brief moment. A slower shutter speed, giving a longer exposure, allows in more light but in that moment of time the subject may move, giving a blurred image. To 'freeze' the motion of a hummingbird's wings, a very fast shutter speed—giving the briefest possible exposure—is vital. A faster shutter speed, however, requires brighter light. Predictably, therefore, capturing images of hummingbirds in flight requires constant and bright low-angled sunlight in order to achieve a high enough shutter speed to freeze the incredibly fast motion of their wings. There are definitely times when the stars align and natural light can be used to capture such images. More often, though, using natural light is an exercise in frustration.

Fortunately, there is one aspect of hummingbirds' behavior that gives us photographers a fighting chance of photographing them in flight. It is simple: hummingbirds are addicted to sugar! They spend a significant portion of their days in pursuing nectar—they have to, in order to fuel their incredibly high metabolism. Because they are so desperate for sugar they can easily be lured to an artificial hummingbird feeder that promises an easy meal. Knowing exactly where a hummingbird is likely to turn up allows photographers simply to set up and wait. This approach does indeed solve part of the challenge of flight photography, but unfortunately the amount and quality of natural light are still unlikely to be adequate for first-rate images.

Booted Racket-tail *Ocreatus underwoodii* | Ecuador

After my first trip to Costa Rica I knew that to capture good-quality images of hummingbirds in flight would require the use of artificial light. If I knew the exact location to which a hummingbird would return over and over again, I would be able to set up a small artificial studio in the field and illuminate the birds properly—the solution to proper illumination being 'multiflash' photography.

Multiflash hummingbird photography works by taking advantage of the incredibly brief pulse of light that is emitted by a camera's external strobe light. By synchronizing several flashes (I typically use four) to light both the bird and the background it is possible to control the lighting conditions completely. The quantity of light that a flash emits is directly related to the duration of the flash bulb's firing. For example, in most camera systems, a full-power (1/1) burst of flash will last for approximately 1/500th of a second. A half-power flash would fire for approximately 1/1,000th of a second and a quarter-power flash would be 1/2,000th of a second. Using this formula, turning the flash down to 1/32 power achieves a flash duration of somewhere around 1/16,000th of a second, which is plenty fast enough to freeze a hummingbird's wings and approximately double the speed of the fastest shutter speed possible when using natural light alone.

There is, however, much more to multiflash photography than simply freezing the hummingbird's motion. That part is in fact very easy to do with the right equipment. Where the skill and artistry come in is in positioning the flashes and background such that the resulting image is beautiful but does not look 'fake'. As I mentioned, I generally use four flashes and they are positioned in such a way as to light both the bird and the background in a way that appears natural. Technically speaking, I could adjust the flashes' distance from the subject to allow for the bird to be completely frozen and for every feather on it to be tack sharp and in perfect focus. The problem with this approach, though, is that it looks so obviously artificial, and it is therefore never my goal. Closing the aperture down as small as possible would give a greater depth of field and the whole bird would look sharply focused. When I am using multiflashes, however, I try to open up the lens as much as possible given the ambient lighting conditions (ideally f/6.3–f/8), so that the depth of field is reduced. After all, when one is trying to photograph hummingbirds in natural light, the aim should be to get the maximum shutter speed possible. This would mean shooting with lenses wide open (perhaps f/4 or f/5.6). My goal with multiflash is to obtain images that look as natural as is possible, and stopping down to more than about f/8 would probably be a move in the wrong direction as it would be unlikely to yield the type of images I am after.

Another part of multiflash hummingbird photography that tends not to get the attention it deserves is that of selecting the right flowers to introduce into an image. By studying the birds I want to photograph in the wild, I know which flowers are the most appropriate to use. I am then able to position these flowers close to the feeder and coat them with a small amount of sugar-water. Instead of including just a small part of the flower in the top of the frame, I try to make the flower as much a part of the composition as the bird itself. Furthermore, to create images that look even more natural, rather than clearly being shot in a studio setting, I try to position the flower so that some parts of the leaves or stem are falling slightly out of focus. When all of these elements are put together well, the resulting images can be nothing short of spectacular.

Aside from the technical aspects of capturing images of hummingbirds, it is also particularly important, I think, to consider the implications for the birds themselves. I have seen photographers take down every single feeder within sight except for the one they are photographing. In my opinion, this is not only unethical but also counterproductive to getting good images. In such cases, what almost always happens is that the most dominant hummingbird will become extremely territorial over that feeder and chase all the others away, leaving the inexperienced photographer wondering where all the birds have gone. Even worse, from an ethical standpoint, are those photographers who actually capture the hummingbirds and place them inside an artificial studio box. This is, to me, unimaginably cruel and should never be done under any circumstances. I always leave multiple food sources available to the birds. If the flash from my camera bothers the birds or stresses them in any way, they can simply choose to visit one of the other feeding stations. The birds must always have the free will to come and go as they please, and should never be forced to visit the feeder at which I am positioned.

Part of what makes the use of multiflash so exciting for the photographer, and so compelling for the viewer, is that it is possible to capture details of something that we are not capable of seeing with our eyes. Hummingbirds simply move too fast for us to appreciate the minute details of their adornments. What we see as a blur of wings and a flash of iridescence can be frozen in time as a perfect moment of pristine glowing feathers—allowing us to admire the beauty of these spectacular birds.

Back home in 'The Lab'

The theory and basic setup of multiflash hummingbird photography is relatively simple to understand and execute. Unfortunately, in the early days of the use of this technique most of the images I was seeing left something to be desired. When I really studied these photographs, I decided that the main problems boiled down to four major flaws. The most obvious problem was that the majority of the backgrounds being used looked completely unnatural. I think that photographers were so happy to be able to freeze the image of the bird that they quite simply put very little thought into the background. A great nature image is the sum of many parts, and the background is undoubtedly as important as the main subject. The second major problem was that, in most of the images I was seeing, the lighting looked very artificial and the subject over-flashed. Problem number three was more of a personal preference than an absolute: almost all of the images I saw were being shot at super-small apertures such as f/22 in order to maximize the depth of field. The expression "just because you can doesn't mean you should" immediately comes to mind: while the power of the flash allows for such settings to be used, these do not necessarily result in a natural-looking image. My final complaint was that very few images celebrated the foliage and flowers as a key part of the picture: flowers were stuck in the corner of the frame as an afterthought and not a central part of the image—such a wasted opportunity!

With these challenges in mind, I set out to try to find a better way of doing multiflash hummingbird photography. Two of the problems (inappropriate use of depth of field through choice of aperture and poor use of flowers) were easy to address in the field. To solve the problems of the backgrounds and lighting, however, I needed to do some research and testing. For the backgrounds, I initially tried using store-bought green fabric stretched out behind my setup. This did not work well at all, as the fabrics were too uniform in color and, if I was not careful when handling them, they would quickly develop wrinkles that would show up in the images. Next, I tried spray-painting thin cardboard in a mix of natural tones, but this was also a complete failure. The answer to ensuring that the backgrounds to my images had a pleasing pallet of natural colors finally turned out to be the making of large prints of out-of-focus photos of appropriate scenes. One important point that I noted, however, was that it was essential for these prints to have a matte finish, otherwise it was quite easy to get glare from the flashes in the image, which was very frustrating.

The challenge of lighting was more difficult to resolve, and I also had to try to figure out a way to carry and set up my 'mini-studio' in the field. This was particularly important as my next photo trip was to be a major one—six months in the tropical Andes, the heart of hummingbird diversity. Because I would be traveling alone and using public transportation, I needed to come up with a method of carrying and setting up a complete multiflash kit by myself. It simply had to be reasonably lightweight and compact.

After a lot of trial and error in my parents' basement, I came up with a system that required six collapsible flash stands. The stands were fairly lightweight (about 1 kg/2.2 lbs each) and easy to set up exactly where I needed them. Four of these stands were to hold the flashes and the flash-receivers. The fifth was to hold my hummingbird feeder and a special articulated arm so that a suitable flower could be 'clamped' in position. The final stand was needed to hold my background, and I designed a specialized aluminum attachment that held a large collapsible light-reflector disk securely in place, to which I could then attach my printed paper. By the end of my research and testing I had a multiflash kit that I could set up anywhere and, while not exactly light at around 12 kg (26 lbs), was at least manageable for me to carry around while chasing hummingbirds throughout the Andes. This is the system I still use today.

FACING PAGE: TOP **Rainbow Starfrontlet** *Coeligena iris* | Ecuador; BOTTOM **Purple-throated Sunangel** *Heliangelus viola* | Ecuador

Notes from the field

Ecuador: the trip of a lifetime and putting theory into practice

Finally the time had come to put all the theory and planning into practice. After completing my Masters degree, I immediately set off on a six-month solo trip to Ecuador for a complete immersion in bird photography. Ecuador is an absolute paradise for birdwatchers and photographers, and is especially rich in hummingbird diversity: over the course of my six months of traveling around the country I was able to photograph more than 60 species of hummingbird. Spectacular species such as Booted Racket-tail, Violet-tailed Sylph (*page 86*) and Velvet-purple Coronet (*page 57*) were just a few of the star birds that I found. But more important than the number of species was the opportunity to develop and perfect my hummingbird-photography techniques. I spent my first month in Ecuador volunteering at Tandayapa Bird Lodge, which has perhaps the best set of hummingbird feeders anywhere. Being able to practice and test out lighting techniques here was absolutely invaluable, and by the time my first month came to an end not only did I have a nice collection of images but I was also getting very confident at photographing these tiny birds.

My Ecuador trip was full of incredible birding adventures. For half a year I chased birds all around the country. While some birds were relatively easy to track down near lodges or hummingbird-feeding stations, others proved to be much more difficult. One species that I was particularly proud of photographing on that trip was the incredibly beautiful Rainbow Starfrontlet. In researching for the trip, I knew that there was basically only one location where these birds were regularly found at a feeding station—Fundación Jocotoco's Utuana Reserve. Unfortunately, the site was not accessible by

public transport and I did not have a car. Luckily for me, when I visited that general area I met a group of birders planning a morning visit to the reserve. They were very clear that they were visiting for only a few hours, so I knew that I would not have much time to work—but at least I had a chance.

On arriving at the site, I was slightly disheartened to find that the location of the feeders was actually quite a distance up a trail. Even worse was the fact that the weather was completely sunny. This would absolutely not work for multiflash, which relies on the flashes to provide 100% of the light in the photo (this is because bright natural or ambient light competes with the light from the flash and results in a double image, one exposed by the flash and the other by the ambient light; this 'ghosting', as it is often called, can be cured only by blocking out all of the natural light from the scene). So, not only would I have to lug all of my gear up the trail and get set up, but I would also have to construct some sort of makeshift shade for the feeder area. I had less than two hours to work!

After a few trips shuttling all of my camera and multiflash gear up to the feeder site, I then got to work on setting up my hummingbird studio. Luckily, I had had lots of practice by this point and had become quite efficient. I then got to work, using a combination of sticks, duct tape, string and a few of my spare background prints to rig up the most rickety-looking shade the world has ever seen. But it worked! Once it was all set up, 30 minutes was all I needed to get some great images of the starfrontlet, and also the bonus of images of Purple-throated Sunangels. I packed everything away and trekked back down the trail, feeling very pleased with myself.

Another great memory from that trip was trying to find and photograph the amazing Ecuadorian Hillstar. This high-elevation specialist is typically found above 4,000 m (13,100 ft) on the slopes of volcanoes. The key to finding and photographing this species is first to locate a good source of their favorite blossoms: Flower of the Andes *Chuquiragua jussieui* flowers. My expedition to photograph this species took me to an area at the base of the still-active Cotopaxi volcano. After driving into the park as far as possible, I scanned the surroundings and eventually spotted a dense area of *Chuquiragua* flowers a few kilometers away on the other side of a large lake. With no other options available, I set off in pursuit of the hillstar.

When I reached my destination I was pleased to find that the Chuquiragua bushes were absolutely loaded with flowers. Sure enough, it was not long before a stunning male Ecuadorian Hillstar emerged and perched atop his prized plant. It would have been the perfect shot—if only the light had not been so brutally harsh.

Sapphire-vented Puffleg *Eriocnemis luciani* | Ecuador

By the time I had made my way to the site, walked around the lake and found the birds it was midday, the sun was directly overhead and the light completely unusable. As is frequently the case with bird photography, once you have located the bird and got close enough to capture an image—not always easy—there is then the matter of dealing with ever-complicated lighting scenarios. This can be incredibly frustrating and time-consuming. My options at this point were simple: either give up (highly unsatisfying) or find a comfortable spot to sit and wait for a cloud (boring). I chose the latter.

As often happens in the mountains, the clouds did eventually begin to build, and a few hours after I first arrived at the site I was finally able to begin trying to photograph the hillstars. Things were going pretty well, with light clouds providing nice diffuse light to work with,

FACING PAGE **Ecuadorian Hillstar** *Oreotrochilus chimborazo* | Ecuador

but in the distance I noticed a dark and ominous sky. At first I did not pay much attention to the approaching storm, but my time in the Rocky Mountains in Canada has taught me that mountain weather can change alarmingly fast. All too quickly the conditions changed from way too sunny, to ideal, to scary. Rain began to hammer down and I started heading back to my car at a brisk pace, spurred along by the increasingly heavy rain. The drops were getting bigger and bigger, and before I knew it I was being pelted by marble-sized hail stones. I made it back to the safety of my car just in time to experience by far the loudest thunderclap of my life. Lightning must have struck the closest hill because the sound boomed through my chest and every hair on my arm stood straight up. It was intense, and a bit scary, but I did not care. After all, I now had my coveted hillstar photos!

Costa Rica revisited

Several years after my first trip to Costa Rica I returned to this wonderful Central American country with a very specific photographic vision in mind. I wanted to try to capture an image of a unique behavior that I had witnessed on that first trip: hummingbirds bathing.

Like all birds, hummingbirds are very fastidious about keeping their feathers clean, constantly preening and often bathing several times a day. Photographing this behavior, however, is next to impossible unless you know exactly where they will do so. Fortunately, I knew of just such a place, as on my first trip to Costa Rica I had been shown a spot at Rancho Naturalista, one of the lodges I had visited, where several species of hummingbird regularly come to a network of small pools to bathe. To this day it is the only site I have ever seen where you can witness this intimate behavior predictably. It is a very special place indeed.

On my first trip I had managed to capture some photos of the hummingbirds bathing—but they were terrible! The site is incredibly dark, so I had used a very high ISO (increasing the sensitivity of the sensor to allow a faster shutter speed, but at the expense of image quality) and a flash (which at the time I did not really know how to operate). Although the photos were dreadful, the image of those tiny birds splashing around in the water was ingrained in my memory. I knew that one day I would return to attempt to photograph them properly.

On my second visit, my plan was to use my new multiflash skills and equipment to freeze the motion

FACING PAGE: TOP **Violet Sabrewing** *Campylopterus hemileucurus* | Costa Rica
BOTTOM **Green Hermit** *Phaethornis guy* | Costa Rica

Purple-crowned Fairy *Heliothryx barroti* | Costa Rica

Green-crowned Brilliant *Heliodoxa jacula* | Costa Rica

of the birds as they erupted from the water. This type of image is both technically and logistically quite difficult to capture. As well as being very dark, the site concerned is in a steep river valley, making the set-up a challenge, to say the least. While dark conditions are not a problem in terms of taking the actual images, as these are illuminated by flash, they are a big problem when it comes to trying to acquire focus. Another challenge in this situation was that the birds are very shy, presumably aware that they are particularly vulnerable to predation while bathing—so the slightest change to their environment or movement by a photographer could frighten them off.

I approached this shoot with patience and determination, and allocated an entire week to give myself a reasonable chance of success. On the first three days I simply visited the site briefly and introduced a few of my flash stands each time. I did not want to make too many changes all at once and risk scaring away the birds. Finally, on days four, five and six, I spent about six hours each day sitting, camouflaged, in a small spot I had selected. The birds would typically come in to bathe about once every hour and would stay for just a few brief seconds. I had to remain very focused on the small bathing pools or I might miss a visit altogether and have to wait again. Even when I was on top of the situation, many times my camera would simply not focus on the bird. This was incredibly frustrating, but in the end all those hours of set-up and waiting did pay off and I came away with a handful of very special images from that week.

Peru: in search of the holy grail

Like all of the tropical Andean nations, Peru has its fair share of hummingbirds, with stunning species such as the Royal Sunangel, White-tufted Sunbeam and Western and Eastern Mountaineers, to name just a few. Without question, however, the hummingbird that most captures the interest of birdwatchers is the Marvelous Spatuletail, a stunning and beautiful bird that is found only in a small area of northern Peru. Sadly, the spatuletail is, like so many species with a restricted range, threatened with extinction (it is categorized as Endangered). Although these are quite small hummingbirds, the male has ridiculously long tail feathers with large 'spatules' at the end. Not only is this a rare species, but it is hard to find and photograph: without question the Marvelous Spatuletail is the holy grail of all hummingbirds.

In 2011, I was on a three-month photo expedition throughout Peru and carved out a few weeks for the northern part of the country. The Marvelous Spatuletail was, of course, my number one target. With a bit of research, it turned out that finding these little beauties

Marvelous Spatuletail *Loddigesia mirabilis* | Peru

was in fact more straightforward than I had thought, as the American Bird Conservancy had been able to partner local conservation organizations in Peru and establish a reserve, the Huembo Reserve, especially for this species. Birders and photographers will forever be in their debt, as protecting habitat for such range-restricted species is absolutely essential to their long-term survival.

Within minutes of arriving at the reserve, I was directed towards a small hillside where a hummingbird feeding station had been established. It was not long before I had my binoculars zeroed in on what has to be one of the world's truly unique birds. But, as we all know, finding a bird and photographing it are two very different things. As is often the case, a patient approach proved to be the most successful. During my first day at the site I observed the birds and took a few record shots. While doing so, I noticed that one exceptionally stunning male spatuletail was very fond of one feeder in particular. The next day I returned and slowly set up my multiflash studio around the feeder of choice, and was able to come away with one of the most favorite bird images I have ever taken. Even more rewarding than the image itself is the fact that it has subsequently been used by conservation groups to promote the reserve and the work that goes in to protecting this incredible bird.

Colombia and Bolivia: targeting endemics

For those of us who become enchanted with birds in the 'New World' tropics there seems to be a natural order regarding the countries that we visit. In many cases Costa Rica is our first experience, often followed by Ecuador and Peru. Before long, though, many Neotropical birders cannot resist a trip to Bolivia or to the mega-biodiverse country of Colombia, which has more species of bird than any other country in the world.

By the time I visited Colombia, I had previously spent a considerable amount of time elsewhere in the tropical Andes and therefore had already photographed many of the more common and widespread hummingbirds. Because of this, I was able to focus my efforts on trying to photograph some of the endemic species, such as Buffy Helmetcrest and Black-backed Thornbill (*page 138*), and other range-restricted species such as Blue-throated Starfrontlet (*page 213*) (which occurs also in neighboring Venezuela). Photographing hummingbirds such as these always demands extra attention on a trip, and the fact that they are so rare most definitely adds an extra element of pressure to the hunt.

As a bird-photographer, you know that there are occasions when you will get just the one chance with a rare or difficult-to-find species—and it can be maddeningly frustrating when nature decides to be

Buffy Helmetcrest *Oxypogon stuebelii* | Colombia

Black-hooded Sunbeam *Aglaeactis pamela* | Bolivia

uncooperative and the subject in question remains elusive. Several years ago, while in Bolivia, I was in pursuit of the particularly beautiful endemic Black-hooded Sunbeam. Adding to the allure of this hunt was the fact that I had never seen a good photo of the species.

My quest for the Black-hooded Sunbeam was focused in central Bolivia, where all my research suggested the bird should be. I searched and searched the known sites for this hummingbird without even catching a glimpse of the target. After several days and at least two flat tires I ran into a birdwatching group, the members of which mentioned that they had seen the sunbeam at 'The Site'. If only I had known about 'The Site' two days before, when I drove right past it!

Retracing my steps back down bumpy Andean roads, I eventually found the spot that had been described to me. All day I searched for this little hummingbird. After a second day, despite my best efforts, three more flat tires and hundreds of kilometers traveled, the best I could manage was a glimpse of my target zipping away into the distance. Reluctantly, I decided to throw in the towel and accept defeat—after all, there are always other birds to chase and there is only a certain amount of time that can be dedicated to just one species.

Sometimes nature will, however, reward your hard work and determination, often when you least expect it. Several days later I was much farther north in Bolivia, near Lake Titicaca. While driving, I noticed a roadside area with a large patch of flowering shrubs. I decided to pull over and investigate, and within five minutes my much-sought-after Black-hooded Sunbeam appeared out of nowhere to fuel up on the plentiful supply of nectar. I had a few minutes to take in the beauty of this bird and capture the best images I could, and then, just as quickly as it had arrived, it zipped off over the hillside never to be seen again. What a gift!

Guyana: targeting a habitat-specific hummingbird

While some hummingbirds, such as the Buffy Helmetcrest and Black-hooded Sunbeam, are restricted in their geographic range, others are more widespread but have very specific habitat requirements. One such species is the stunning Crimson Topaz, which occurs in lowland rainforest mainly in eastern Venezuela, Guyana, Suriname, French Guiana, and northern Amazonian Brazil. Unfortunately, it rarely visits feeders and tends to be seen most frequently in the canopy of flowering trees, not exactly an ideal situation for photography! All of the field guides, however, mention that the topazes (of which two species are recognized in the genus *Topaza*, the other being the equally stunning Fiery Topaz that is found across much of the western half of the Amazon Basin) also habitually spend a lot of time around small 'blackwater' streams, perhaps supplementing their diet of nectar with insects that are easiest to see and catch over open water. In order to photograph this species, I would have to find an accessible stream surrounded by suitable lowland rainforest habitat and then, as is often the case, wait patiently.

My opportunity to photograph a Crimson Topaz finally came while I was deep in the rainforest of Guyana. A local guide had told me that he regularly saw these shimmering red beauties around a stream that was just a kilometer or so from Atta Lodge, where I was staying. After a sweaty hike through the sweltering jungle to get to the site, I decided that my best approach was to position myself in photographic range of some obvious twigs that were extending out over the stream. To increase my chances of success I positioned a small Bluetooth speaker close to the overhanging perches, the idea being that, if I heard or saw the topaz in the area, I could potentially lure the bird to my target zone by using playback.

It is such a thrill when a lot of hard work, research, knowledge of the subject's behavior, patience, and of course a healthy dose of luck, pay off and result in the type of image that I had dreamt of. As a bird-photographer, it is so exciting when, having stared at a beautiful bird on the pages of field guides for many years, one finally gets to see it for real, particularly when the bird itself is even more beautiful than any illustration on a page could ever convey. This was certainly the case with the Crimson Topaz! Not even John Gould's gold-leaf lithographic prints (see *page 184*) could accurately depict the stunning plumage of this bird, and, while my photos, too, will surely fall short of the experience of actually seeing such birds in the wild in their natural habitat, perhaps they are the next best thing.

Crimson Topaz *Topaza pella* | Guyana

Jamaica and Brazil: easy and unusual

Photographing hummingbirds such as the Crimson Topaz or Marvelous Spatuletail is always a challenging endeavor, and as bird-photographers we come to expect and enjoy the thrill of the hunt. But I will be the first to admit that photography can definitely be far more pleasurable on the rare occasions when the birds make it easy. On a recent trip to Jamaica, my number one most-wanted bird for the entire trip was the Black-billed Streamertail. This endemic hummingbird is restricted to the northeastern part of Jamaica, and I was uncertain of how difficult it would be to track it down. To my delight, it turned out that the hotel at which I was staying had a hummingbird-feeder in the bar. Within minutes of arriving I was treated to my first looks at this spectacularly adorned little hummingbird. The next day, during the 'off-peak' drinking hours for the local human patrons, I negotiated with the hotel management to let me set up my multiflash equipment in the bar, and before long I was looking at the shots of my dreams on the back of my camera. It sure is nice when things are easy for a change!

Although photographing streamertails while enjoying a cocktail was definitely far from the norm for me, it was by no means the most unusual experience I have had when taking pictures of hummingbirds. That honor goes hands down to the Stripe-breasted Starthroat that I photographed in Brazil. My quest for this bird took me to the small town of Vargem Bonita, in central Brazil. Trip reports and internet research had led me here, as they mentioned a local resident who had a hummingbird-feeder set up in his backyard that was attracting this special hummingbird. After asking a few of the locals, I eventually found the house. The owner graciously invited me in to his small urban backyard where, sure enough, there was a hummingbird-feeder and my target hummingbird. I explained to the owner what I was hoping to achieve and he was kind enough to allow me to set up my equipment. All was going more or less according to plan when the owner appeared from inside his house with an accordion. For the next hour I had one of the strangest, most charming and memorable of all my photographic experiences: the starthroat 'performed' beautifully while I listened to the owner belting out Portuguese melodies and tickling the ivories on his squeeze-box. This just goes to show that you never know where the quest for birds will take you!

Final thoughts: a worthy pursuit

I love hummingbirds for their beauty, for their acrobatic abilities, and for the ingenious ways in which they have evolved to solve problems over millions of years. But even more than these characteristics, I love hummingbirds for the places they have led me to throughout the Neotropics. Bird photography for me is about many things. It is about creating stunning images to share with people around the world. It is about achieving a hard-fought goal. It is about being outside and using all of my senses to explore and understand the natural world in a more profound way. Perhaps more than any of these tangible outcomes, however, bird photography becomes the excuse to visit wonderful and unusual places that I would otherwise never have found.

I hope that the images of hummingbirds in this book will engender a deeper appreciation of these extraordinary creatures and the fascinating lives they lead. Perhaps they will encourage you to seek out some of these amazing birds yourself and find an adventure of your own along the way. In doing so, you will undoubtedly help hummingbirds in some way. Whether this is simply through increasing awareness of the brilliance of hummingbirds in general, or by contributing your tourism spending directly to a conservation initiative, it all helps. After all, ecotourism in its many forms, particularly where it involves visits to lodges in far-flung places and traveling with local guides, does much to help support local communities and maintain protected habitats.

It is painful for me to think of a world with fewer species of hummingbird. Sadly, such a fate seems inevitable in the near future if conservation and habitat-protection measures are not ramped up dramatically. Anything we can do to prevent such a tragedy is surely worthwhile, making our pursuit of hummingbirds, whether with binoculars and a notebook or a camera, all the more satisfying.

FACING PAGE **Black-billed Streamertail**
Trochilus scitulus | Jamaica

BELOW **Stripe-breasted Starthroat**
Heliomaster squamosus | Brazil

In Search of Nature's Jewels: A Personal Quest

Andy Swash

Starting young

Ever since the age of about seven, when my parents took me to the Natural History Museum in London and stood by while I looked in awe at the Victorian hummingbird display cabinet in the Bird Gallery, I have been hooked on these amazing birds. Although my parents did not have any great interest in the subject themselves, they were remarkably supportive of a young boy with an insatiable curiosity for all things 'natural'—and so it was that my interest in wildlife was allowed to flourish.

I spent my childhood in the south of England in the 1960s and, with abundant wildlife on my doorstep and the freedom to explore, I had the ideal opportunity to get to know the local area very well. I was particularly keen on birds and, thanks to a wonderful mentor living nearby, and the encouragement of teachers at my school, I became quite proficient at identifying and recording what I saw. In my late teens I started to undertake voluntary conservation work on local nature reserves, and the experience I gained led to my first proper job—as assistant warden at Brownsea Island Nature Reserve, in Dorset.

In the following year I was very fortuitously selected to study zoology at Manchester University, at the time one of only a handful of universities that offered ornithology as a specific study option. Thanks to my course tutor, who recognized my particular interest in birds, I was granted access to the Dresser Collection of bird skins (one of the largest such collections in the world), housed at the Manchester Museum. It will perhaps come as no great surprise that it was the drawers of hummingbirds that I was particularly interested in examining—and in the process reigniting the wonderment I had felt some 12 years earlier when visiting London's Natural History Museum. As part of my studies, I undertook a research project into feather microstructure and started to gain some understanding of how iridescence works. This was all fascinating stuff, but after looking for so long at so many hummingbird skins, and dreaming about the places from which they had originated, I was determined one day to visit the Americas to see a live one for myself. I would have to wait for a few years, however, before I was able to save up enough money to fulfil my dream.

A dream comes true

Saturday April 4, 1987 is still vividly ingrained in my mind. By this time I was married, and my wife, Gill, and I were at the beginning of our first birding trip to the Americas—a month-long journey that was to take us from Florida, through the southern states and right across to California. One of the first sites we visited was the Arthur R. Marshall Loxahatchee National Wildlife Refuge and, as we walked from the parking lot to the visitor center, a slight movement around some flowering plants nearby caught my eye. I raised my binoculars and stood transfixed as there, right

Ruby-throated Hummingbird *Archilochus colubris* | USA

FACING PAGE **Long-tailed Sylph** *Aglaiocercus kingii* | Colombia in front of me, was a beautiful male Ruby-throated Hummingbird with his gorget flashing red as he zipped from flower to flower. After collecting its precious nectar, the bird hovered briefly, flashed its gorget once more and zoomed off into the distance, never to be seen again. That was it, a dream fulfilled! But the impact that little sprite had on me was quite profound, as I now really was hooked on hummingbirds and just wanted to see more....

By the time we boarded the plane home after that trip we had seen 11 species of hummingbird, a few of which I had also managed to photograph. Admittedly, the photos were not great, but they were good enough to provide a lasting memento of our first Magnificent, Blue-throated, Black-chinned, Costa's, Broad-tailed, Broad-billed, Calliope, Rufous, Allen's and Anna's Hummingbirds in the deserts, mountains and canyonlands of Arizona, Utah and California. Two years later we returned to the United States and spent a few weeks birding in Texas, where we added another two species to our list—Buff-bellied and Lucifer Hummingbirds—and also spent countless hours watching and admiring many of the species that we had seen previously. Having now virtually exhausted the hummingbird options in the USA, though, we would have to travel elsewhere if we wanted to see more species. And so it was that, in 1991, we began what has become a pretty much annual pilgrimage to tropical America in search of our favorite avian jewels.

Continuing the quest—Central America

One of the hummingbird hotspots of Central America is the easily accessible country of Costa Rica, and this was to be the next port of call in our quest. After a month exploring the tremendous variety of habitats, from coastal mangroves to rainforest-shrouded mountains and high volcanoes with *páramo*-type vegetation, we had seen no fewer than 40 different species of hummingbird with an array of fantastic names: ranging from hermits and Band-tailed Barbthroat to Green-fronted Lancebill, violet-ears, Purple-crowned Fairy, Green-breasted Mango, Green Thorntail, White-crested Coquette, sabrewings, mountain-gems and the elusive Snowcap, to name just a few. The feeders at Monteverde Cloud Forest Reserve certainly lived up to their reputation as being one of the best places to watch, enjoy and become familiar with a wide variety of hummingbirds—slowly learning the distinguishing features of the various species, particularly the often confusingly similar females, all being part of the fun. And it has to be said that doing so during the occasional earth tremor certainly added a frisson of excitement to the experience!

Snowcap *Microchera albocoronata* | Costa Rica

South America at last!

In the following year, Gill and I visited Venezuela and spent three weeks driving a loop from Caracas to Henri Pittier National Park just to the west, and on to Mérida in the Andes and back via the vast low-lying plains of the Orinoco Basin, los llanos. During this journey, which took us from tropical forests through desert to high *páramo* and extensive wetlands, we saw 34 species of hummingbird—only a handful of which we had seen previously in Costa Rica. Our first taste of 'Andean' hummingbirds was a particular highlight, and the sightings we had of Long-tailed Sylph, Glowing Puffleg and Merida Sunangel will live long in the memory. One of the hummingbirds we most wanted to see was the (now named) White-bearded Helmetcrest. But try as we might, despite spending days scouring the *páramo* for flowering frailejónes—*Espeletia* plants—which are the favored nectar source for the helmetcrest, we drew a complete blank. We would have to return one day if we wanted to see this localized species....

Northern Argentina was on our agenda for the following January, and this gave us the opportunity to enjoy the delights of yet more Andean hummingbirds. The area around Cachi proved to be a great place to see the stunning Red-tailed Comet and the amazing Giant Hummingbird, including, with some good fortune, a bird on a nest that happened to catch my eye when I was scanning a scrubby gulley. The nest was situated halfway up a gnarled, lichen-encrusted branch about 4 meters (12 ft) from the ground. After a considerable amount of scrambling I managed to climb the slope beyond the nest and get into a great position where I could sit and watch the bird at eye level.

The light was fantastic in the clear Andean air, and I was keen to try to get a photo of the scene that I could show to others on my return home. This was, of course, in the days of film, before digital photography was even an option. In those days, eking out our meagre supplies of Kodachrome 64 over the course of a trip was always tricky, especially as we had no idea what photographic opportunities might lie ahead. We had therefore set ourselves a limit of six shots for any such situation—which was a tremendous incentive to ensure that all the camera settings had been checked and that the necessary exposure corrections had been made before pressing the shutter! We then had to wait until we were back at home and the film had been processed before seeing the results, hoping all the while that it had not been fogged by a mis-correlated airport X-ray scanner, and that it was correctly processed and not scratched in any way. Such were the challenges of growing up

Black-breasted Puffleg *Eriocnemis nigrivestis* ♀ | Ecuador

with film cameras as the only option! On this occasion, the stars aligned and resulted in two images that were subsequently published in some prestigious books.

To this day, that beautifully constructed Giant Hummingbird nest is the only one I have ever seen. Thinking back now to the time I spent watching it, I remember being filled with curiosity and wonder. From where had the female got all the material to build her nest and how had she managed to carry it? How long had she taken to build it? How many eggs had been laid? How long would she be incubating? What would the chick(s) be fed on, and how long would they take to fledge? These were just a few of the multitude of questions in my head, which left me with the stark realization that, other than recognizing the bird as a Giant Hummingbird and being aware that only females were involved in nest-building and rearing the young, I knew virtually nothing about its biology or ecology. How long did the young take to become independent? What was the male doing while the female was bringing up their offspring? Were the birds resident or did they move elsewhere at different times of the year? I think it was at that moment that I recognized I really needed to know more about hummingbirds in order to appreciate and enjoy these remarkable birds to the full. More background reading and more research were clearly needed before any future trip!

Our next visit to the Americas was to the vast country of Brazil, where during the early part of the trip we spent many hours watching the hummingbird-feeders at the

ABOVE a female **Giant Hummingbird** *Patagona gigas* on its nest near Cachi, Argentina; BELOW the author taking the photo.

hotels in Itatiaia National Park, a short drive from Rio de Janeiro, in the southeast of the country. This provided an opportunity to get to know the regular hummingbirds of the area, species such as Brazilian Ruby, White-throated Hummingbird, Black Jacobin, Versicolored Emerald and Violet-capped Woodnymph. Although we enjoyed watching these birds' antics at the feeders throughout the day, activity was always particularly frenzied towards dusk when they were stoking up in readiness for the cool night ahead. Visiting Itatiaia also gave us a chance of finding one of the hummingbirds we most wanted to see, the enigmatic Plovercrest. These birds do not visit feeders, and the advice we got from a local birder was to head for the mountains and look for them in the higher parts of the Park.

A bumpy drive up the Agulhas Negras Road brought us to an area of upper montane forest that was dripping with epiphytic lichens and bromeliads, and here our quest began in earnest. We had just started to walk slowly along the road when I heard an unfamiliar rather high-pitched, disyllabic call, somewhat reminiscent of a 'squeaky toy', coming from just inside the forest. Having done my homework this time, I quickly realized that this distinct sound must be our target species, a displaying male Plovercrest! After a few minutes, the bird stopped calling and suddenly there it was, hovering right in front of us, feeding low down from some small red flowers, glistening green and purple. Amazingly, after a few seconds the bird perched briefly on a twig, facing us, breast puffed out and long crest raised like an aerial,

ABOVE **Green-crowned Plovercrest** *Stephanoxis lalandi* | Brazil
'The Plovercrest' is now treated as two species: Green-crowned Plovercrest and Violet-crowned Plovercrest, which, broadly speaking, occur to the east and west respectively of São Paulo;
BELOW Agulhas Negras, Rio de Janeiro, Brazil

before whizzing back into the forest. It was almost as though the bird knew how much we wanted to see it and was simply showing off. It was a magical moment and certainly an unforgettable experience.

The joys of 'Hummingbird Central'

The center of the universe for hummingbirds is the northwest of South America, Ecuador and Colombia between them hosting an astonishing 186 species (just over half of all the world's species). That there are so many hummingbirds in this part of the Americas is due to the great diversity of habitats that occur from sea level up the west slope of the Andes to the lofty peaks and down the east slope to the Amazon Basin. This, then, is the region to visit in order to appreciate and enjoy to the full the tremendous diversity of this amazing bird family. Since our first visit, in the mid-1990s, Gill and I have returned on numerous occasions to enjoy the delights of these two countries.

The great joy of visiting 'Andean' countries if you are hooked on hummingbirds is that there are many wildlife lodges in excellent locations, most of which have feeders that attract a wide range of species. In order to hone our identification skills, we have found that the best strategy is to spend a few days at a range of lodges at different altitudes on both the west and east slopes of the Andes. Next we locate the feeders, and simply sit or stand and marvel at the incredible variety of these tiny birds. The challenge then is that of putting names to the hummingbirds, a process that is frequently hampered by the bewildering amount of activity at some feeders, with birds zipping about in all directions at great speed and visiting to feed only fleetingly. In our experience, however, perseverance usually pays off, as any one bird will keep returning, often on a regular basis, to the same feeder.

When Gill and I first visited Ecuador, we explored on our own, which was always exciting and rewarding but at times could be very frustrating, too, as we were obviously missing so much. Increasingly, we came to realize the benefits of traveling with a local guide who really knew the local area and where, when and how to look for certain species. The other advantages of engaging the services of a local guide, we have found, is that he/she invariably has a deep knowledge of the area and its special wildlife, and is also very often friendly with local farmers and landowners who are willing to allow access to otherwise inaccessible places.

Although spending time at lodge feeders is the best way of seeing many hummingbirds, there are some species that can be seen dependably only by visiting feeders that are sited in remote locations, and in some cases only during certain months of the year. For example, the best chance of seeing the extremely localized and highly threatened Black-breasted Puffleg is to visit Fundación Jocotoco's Yanacocha Reserve near Quito, in Ecuador—and particularly during the period from April to July, when birds will occasionally visit the feeders that are located a fair distance into the reserve.

Some hummingbirds make birdwatching even more exciting, as they are not known to visit feeders at all and can be seen only by exploring areas where their favored foodplants grow. The two species of sicklebill—White-tipped Sicklebill and Buff-tailed Sicklebill—are a prime example. These hummingbirds have a remarkably downcurved bill that has evolved to enable them to take nectar from plants with particularly long, curved, tubular flowers such as *Heliconia* and *Centropogon*, and, as a consequence, are unable to feed from standard hummingbird-feeders. Fortunately, though, *Heliconia* are frequently found growing in lodge grounds and it is therefore sometimes possible to see these amazing birds without having to venture very far.

Over the years that we have spent in searching for hummingbirds, not only in Ecuador and Colombia but throughout the Americas, we have met many wonderful local guides and dedicated conservationists, and made some great friends. Although we do have reservations over our carbon footprint, despite being able to offset it to some extent, one thing that has become clear to us is the importance of ecotourism in maintaining the livelihoods of local people and ensuring that the many excellent wildlife lodges are able to stay open. The overall contribution that this makes in helping to safeguard and conserve so many very special places should not be underestimated.

Seeing a wide variety of hummingbirds at close quarters in lodge grounds or at isolated feeders is certainly an enjoyable and educational experience, but the thrill of a specially planned visit to a remote area to see a particular hummingbird in its true domain is one of the great joys of birding in the Americas. Over the years, Gill and I have traveled to many fantastic places in search of very localized and little-known hummingbirds, often in the hope of taking photos that can be used by local or national conservation organizations—or indeed BirdLife International—to promote their vital work. A number of these photos are featured in the chapter *Conservation: Hummingbirds Under Threat* (*page 191*) and elsewhere in this book.

For me, however, a hummingbird does not have to be rare to engender excitement, as seeing any hummingbird always brings a feeling of joy and wonderment. Watching their antics and being privileged to be in their environment are all part of the enjoyment of hummingbirding (if there is such a word!). In fact, an isolated cactus-strewn rocky plateau near the town of Boa Nova in Bahia, Brazil, is the site of one of the most

FACING PAGE **White-tipped Sicklebill** *Eutoxeres aquila* | Ecuador

remarkable hummingbird spectacles I have ever seen. At certain times of the year, when the barrel cacti are flowering, hundreds of hummingbirds can sometimes gather here, particularly just before dusk to stoke up for the night ahead. One of the species concerned is the stunning Ruby-topaz Hummingbird, and the sight of dozens of these little beauties flying around your feet, seemingly oblivious of your presence, is certainly a sight to behold. I am not a poet, but I did feel compelled to write a verse in my notebook that evening when reliving the wonderful experience and trying to encapsulate it in a few words:

Shimmering gold in the gathering gloom,
skimming and dancing from bloom to bloom.

Final thoughts: making a difference

I am very fortunate to have been able to travel widely in the Americas and to have had the opportunity to observe and photograph over three-quarters of the world's hummingbird species. The photos shown here, and those used to illustrate other chapters of this book, are some of my personal highlights from these travels over the past two decades since the advent of digital cameras. The experience gained from seeing and studying such a large number of species, and particularly many of those that are highly threatened, has increasingly left me wondering what I can do personally to help safeguard the future of these incredible birds. Apart from making a financial contribution to conservation initiatives, I have always had the ambition to share my experience and knowledge of the various aspects of the biology and ecology of hummingbirds, their cultural significance and the threats that they face, by presenting these in a form that others can benefit from and appreciate. That was my personal motivation for wanting to produce a book about hummingbirds, and meeting Glenn at Eco-lodge Itororó in Brazil back in 2017, entranced by the many species visiting the feeders, was the catalyst.

ABOVE **Frilled Coquette** *Lophornis magnificus* | Brazil

FACING PAGE **Ruby-topaz Hummingbird** *Chrysolampis mosquitus* | Brazil

This rocky plateau with its flowering barrel cacti provided the backdrop for an amazing spectacle of dozens of hummingbirds feeding at dusk. | Boa Nova, Bahia, Brazil

Taxonomy: The BirdLife List of Species

Taxonomic conundrums

Taxonomy is the branch of biology that deals with describing, naming and classifying groups of organisms based on shared characteristics. Serious scientific attempts to apply this to birds date back to the 17th century, while global bird lists, developed initially as catalogs to facilitate the organization of bird collections, emerged at the turn of the 20th century. The *Peters' Check-list* (published 1931–1987) is arguably the best known of these, and is still influential, not least for its attempts to consolidate the number of bird species by grouping together (lumping) similar forms.

Until the latter half of the 20th century, decisions were made exclusively by analyzing the way a bird looks (known as its morphology). In particular, ornithologists looked at aspects of plumage and skeletal structure to classify birds into families, genera and species. The advent of biochemical analyses, which became increasingly influential in both defining species and determining higher-order relationships, revolutionized bird classification, frequently producing surprising results. Advances in molecular analysis, and to a lesser extent the recognition of vocal characters in defining species, have permitted major advances in the understanding of bird taxonomy over the past few decades, and resulted in significant changes to bird checklists and names. At the same time, the rapid advances have encouraged different approaches to interpreting novel data, and arguments over, and frustration with, bird names have become a feature of both professional and amateur

FACING PAGE The **Magnificent Hummingbird** *Eugenes fulgens* orrurs from southwest USA, through central America to west Panama. It is treated as a single species by BirdLife, but split into two species by other taxonomic authorities (see note [42] on *page 270*). | Panama

BELOW The recently described **Dry-forest Sabrewing** *Campylopterus calcirupicola* (which some taxonomists refer to as Outcrop Sabrewing) looks very similar to, and is clearly a close relative of, the Grey-breasted Sabrewing (see *page 165*). It is, however, geographically isolated and is restricted to drier forests in the interior of eastern Brazil. Perhaps uniquely, it nests in caves. Amazingly, another isolated population formerly considered to be a subspecies of Grey-breasted Sabrewing occurs just to the south of the range of Dry-forest Sabrewing, and it, too, has recently been recognized as a separate species, Diamantina Sabrewing *Campylopterus diamantinensis*. To complicate matters further, the population of Grey-breasted Sabrewing that occurs in eastern Amazonia may prove to be yet another separate species once further studies have been undertaken. This just goes to illustrate the current state of flux with hummingbird taxonomy, and it seems inevitable that there will be further such revelations in the future. | Brazil

Hummingbird phylogeny

The most recent scientific studies suggest that hummingbirds (family Trochilidae) diverged from swifts (family Apodidae) and treeswifts (family Hemiprocnidae) during the Eocene geological epoch some 40 million years ago (mya), and started to diversify within themselves around 22 mya. This is summarized in the graphic interpretation of the evolutionary relationships between hummingbirds and their closest relatives that is shown on *page 13*.

The graphic below is an alternative presentation of the 'family tree' for just hummingbirds. As well as showing the approximate period when each of the subfamilies is believed to have diverged from a common ancestor, it provides an indication of how the relative number of species within each subfamily has increased as time has progressed since then. The color-coding used for the subfamilies and clades is applied in this section, which covers the BirdLife list of species.

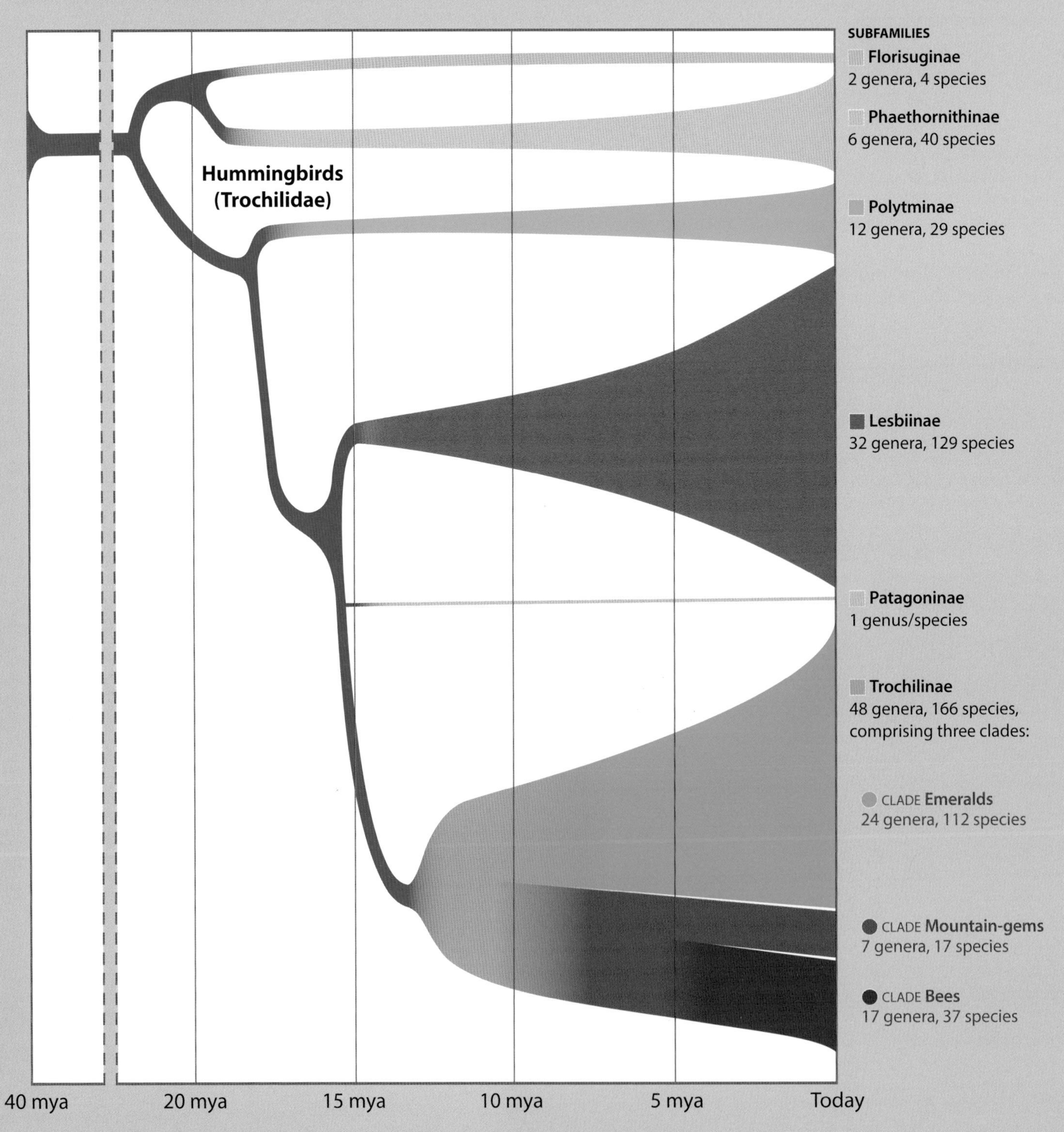

ornithology. As a result, there are four major world checklists of birds that are widely used today: *The Howard and Moore Complete Checklist of the Birds of the World*; *IOC* (International Ornithological Community) *World Bird List*; *eBird/Clements Checklist*; and *HBW* (Handbook of the Birds of the World) *and BirdLife International* (see *Further Reading and Sources of Useful Information, page 277*).

The adoption of one or another bird taxonomy has implications beyond the mere naming of birds. Indeed, it may have consequences for the very survival of the birds themselves. Putting a name to something acknowledges its existence. It is often the case that only by putting a name to a species can it gain a status that merits study—and only then can it be cherished, cared for and appreciated. For example, Guanacaste Hummingbird (see *page 197* in the *Conservation* chapter) is recognized as a species only under *HBW and BirdLife International* taxonomy, and is assessed as Critically Endangered. Such recognition immediately raises its public profile, facilitating fund-raising and therefore work to conserve the species, something that has triggered ongoing searches of the type locality and likely range. The situation regarding the recently discovered Blue-throated Hillstar is rather more straightforward given its acceptance as a new species by all taxonomic lists—but most importantly by BirdLife International as the IUCN Red List Authority. Under these circumstances, efforts to ensure the survival of a highly threatened and very localized bird, such as this beautiful hummingbird, can be assured. Since the *HBW and BirdLife International* checklist is used by BirdLife International, and adopted by the International Union for Conservation of Nature (IUCN) and several international treaties, it is the *de facto* taxonomy used by conservation organizations, and therefore the choice for this present book.

The **Black-throated Brilliant** *Heliodoxa schreibersii* is one of many taxonomic conundrums among hummingbirds. Isolated populations in central and southeast Peru are treated by BirdLife as a separate species, Black-breasted Brilliant *H. whitelyana*, whereas IOC, eBird and H&M currently treat the populations as a single species. | Ecuador

The BirdLife List of Species

BirdLife International currently recognizes 101 genera, 369 extant species and 714 taxa (species and subspecies) of hummingbirds. It is beyond the scope of this book to attempt to illustrate all the species in their various plumages—to do this justice would require another volume—but the following pages provide an overview of the family and illustrate the range and variation of forms and plumages.

An example of at least one representative of every hummingbird genus is shown in this section, and the tabulated information that follows (from *page 256*) lists all the species, cross-referenced, where appropriate, to pages in the book where a photograph appears (in total, 262 species are illustrated—more than 70% of all the world's hummingbirds). The entry for each species is based on the latest BirdLife data and covers the following: 2021 IUCN global Red List status (see *pages 191–192*), population trend, number of subspecies recognized (species with no subspecies are termed monotypic), and distribution (broad summary and map). Details are also included on the species' altitudinal breeding range, based on information gleaned from various publications; since many hummingbirds are altitudinal migrants, there may be outlying records from higher or lower elevations. The coding used in the tabulated list of species is explained at the bottom of *page 255*.

Up-to-date information on the species, including its status, distribution, population estimates and threats, is available on the BirdLife International website at http://datazone.birdlife.org/home (which can be accessed via the quick response code [QR code] *above* using a QR reader that can be downloaded as an app for smartphones or tablet computers.

This overview of the genera and species is followed by two lists of the hummingbirds that have previously been described as species but are now considered to be invalid. The first list covers 30 hummingbirds that are now known to be the same as another named species, or which are suspected, or have been shown, to be aberrant or immature forms of other species (*Appendix 1, page 274*). The second lists 34 hummingbirds that have been described as separate species but are now known to be hybrids (*Appendix 2, page 275*).

Subfamily **Florisuginae**

Topaza 2 species | *p. 256*
Crimson Topaz *T. pella* | Guyana

Florisuga 2 species | *p. 256*
White-necked Jacobin *F. mellivora* | Colombia

Subfamily **Phaethornithinae**

Eutoxerxes 2 species | *p. 256*
White-tipped Sicklebill *E. aquila* | Ecuador

Ramphodon 1 species | *p. 256*
Saw-billed Hermit *R. naevius* ♀ | Brazil

Glaucis 3 species | *p. 256*
Rufous-breasted Hermit *G. hirsutus* | Brazil

Threnetes 2 species | *p. 256*
Band-tailed Barbthroat *T. ruckeri* | Guyana

Anopetia 1 species | *p. 256*
Broad-tipped Hermit *A. gounellei* | Brazil

Phaethornis 30 species | *p. 256*
Scale-throated Hermit *P. eurynome* | Brazil

Subfamily **Polytminae** 1/2

Doryfera 2 species | *p. 258*
Green-fronted Lancebill
D. ludovicae | Colombia

Schistes 2 species | *p. 258*
Western Wedge-billed Hummingbird
S. albogularis | Colombia

Augastes 2 species | *p. 258*
Hyacinth Visorbearer
A. scutatus | Brazil

Colibri 4 species | *p. 258*
White-vented Violet-ear
C. serrirostris | Brazil

Androdon 1 species | *p. 258*
Tooth-billed Hummingbird
A. aequatorialis | Colombia

Heliactin 1 species | *p. 258*
Horned Sungem
H. bilophus | Brazil

Heliothryx 2 species | *p. 258*
Purple-crowned Fairy *H. barroti* | Colombia

Polytmus 3 species | *p. 258*
White-tailed Goldenthroat *P. guainumbi* | Brazil

Chrysolampis 1 species | *p. 258*
Ruby-topaz Hummingbird *C. mosquitus* | Brazil

Subfamily **Polytminae** 2/2

Avocettula 1 species | *p. 258*
Fiery-tailed Awlbill
A. recurvirostris | Brazil

Anthracothorax 8 species | *p. 258*
Black-throated Mango *A. nigricollis* | Brazil

Eulampis 2 species | *p. 259*
Purple-throated Carib
Eulampis jugularis | Martinique

Subfamily **Lesbiinae**

Heliangelus 9 species | *p. 259*
Amethyst-throated Sunangel
H. amethysticollis | Ecuador

Sephanoides 2 species | *p. 259*
Green-backed Firecrown
S. sephaniodes | Chile

Discosura 2 species | *p. 260*
Green Thorntail *D. conversii* | Ecuador

Lophornis 11 species | *p. 260*
Spangled Coquette *L. stictolophus* | Ecuador

Phlogophilus 2 species | *p. 260*
Ecuadorian Piedtail
P. hemileucurus | Ecuador

Adelomyia 1 species | *p. 260*
Speckled Hummingbird
A. melanogenys | Ecuador

Aglaiocercus 1 species | *p.260*
Long-tailed Sylph *A. kingii* | Ecuador

Sappho 2 species | *p.260*
Red-tailed Comet *S. sparganurus* | Bolivia

Taphrolesbia 1 species | *p.260*
Grey-bellied Comet *T. griseiventris* | Peru

Polyonymus 1 species | *p.261*
Bronze-tailed Comet *P. caroli* | Peru

Oreotrochilus 7 species | *p.261*
Ecuadorian Hillstar *O. chimborazo* | Ecuador

Opisthoprora 1 species | *p.261*
Mountain Avocetbill *O. euryptera* | Colombia

Subfamily **Lesbiinae**

Lesbia 2 species | *p. 261*
Green-tailed Trainbearer
L. nuna | Ecuador

Ramphomicron 2 species | *p. 261*
Purple-backed Thornbill
R. microrhynchum | Colombia

Chalcostigma 5 species | *p. 261*
Rufous-capped Thornbill
C. ruficeps | Ecuador

Oxypogon 4 species | *p. 261*
Green-bearded Helmetcrest
O. guerinii | Colombia

Oreonympha 2 species | *p. 262*
Eastern Mountaineer
O. nobilis | Peru

Metallura 9 species | *page 262*
Viridian Metaltail *M. williami* | Ecuador

Haplophaedia 3 species | *p. 262*
Greenish Puffleg *H. aureliae* | Colombia

Eriocnemis 12 species | *p. 262*
Golden-breasted Puffleg *E. mosquera* | Ecuador

2/3

Loddigesia 1 species | *p. 263*

Marvelous Spatuletail *L. mirabilis* | Peru

Aglaeactis 4 species | *p. 263*

White-tufted Sunbeam *A. castelnaudii* | Peru

Coeligena 19 species | *p. 263*

Bronzy Inca *C. coeligena* | Ecuador

Lafresnaya 1 species | *p. 264*

Mountain Velvetbreast *L. lafresnayi* | Colombia

Ensifera 1 species | *p. 264*

Sword-billed Hummingbird *E. ensifera* | Ecuador

Pterophanes 1 species | *p. 264*

Great Sapphirewing *P. cyanopterus* | Colombia

Subfamily **Lesbiinae** 3/3

Boissonneaua 3 species | *p. 264*
Buff-tailed Coronet *B. flavescens* | Ecuador

Ocreatus

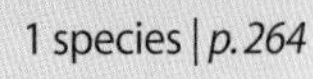

1 species | *p. 264*
Booted Racket-tail *O. underwoodii* | Ecuador

Urochroa 2 species | *p. 264*
White-tailed Hillstar *U. leucura* | Ecuador

Urosticte 2 species | *p. 264*
Rufous-vented Whitetip *U. ruficrissa* | Ecuador

Heliodoxa 10 species | *p. 264*
Violet-fronted Brilliant *H. leadbeateri* | Ecuador

Clytolaema 1 species | *p. 265*
Brazilian Ruby *C. rubricauda* | Brazil

Subfamily **Patagoninae**

Patagona 1 species | *p. 265*
Giant Hummingbird *P. gigas* | Ecuador

The **Giant Hummingbird** is by far the largest of all hummingbirds, and is the sole representative of the subfamily Patagoninae.

Subfamily **Trochilinae** Emeralds ; Mountain-gems ; Bees **1/4**

Chlorostilbon 15 species {+2 extinct} | *p. 265*
Glittering-bellied Emerald *C. lucidus* | Brazil

Cynanthus 4 species | *p. 266*
Broad-billed Hummingbird *C. latirostris* | Mexico

Cyanophaia 1 species | *p. 266*
Blue-headed Hummingbird
C. bicolor | Martinique

Klais 1 species | *p. 266*
Violet-headed Hummingbird
K. guimeti | Costa Rica

Abeillia 1 species | *p. 171*
Emerald-chinned Hummingbird
A. abeillei | Mexico

Orthorhyncus 1 species | *p. 266*
Antillean Crested Hummingbird
O. cristatus | Lesser Antilles

Subfamily **Trochilinae** Emeralds ; Mountain-gems ; Bees

Stephanoxis 2 species | *p. 266*
Green-crowned Plovercrest *S. lalandi* | Brazil

Aphantochroa 1 species | *p. 266*
Sombre Hummingbird *A. cirrochloris* | Brazil

Anthocephala 2 species | *p. 266*
Tolima Blossomcrown *A. berlepschi* | Colombia

Campylopterus 13 species | *p. 266*
Lazuline Sabrewing *C. falcatus* | Colombia

Eupetomena 1 species | *p. 267*
Swallow-tailed Hummingbird *E. macroura* | Brazil

Eupherusa 4 species | *p. 267*
Stripe-tailed Hummingbird *E. eximia* | Costa Rica

Elvira 2 species | *p. 267*
Coppery-headed Emerald
E. cupreiceps | Costa Rica

Microchera 1 species | *p. 267*
Snowcap
M. albocoronata | Costa Rica

Chalybura 2 species | *p. 267*
White-vented Plumeleteer
C. buffonii | Colombia

2/4

Thalurania 5 species | *p. 268*
Violet-capped Woodnymph
T. glaucopis | Brazil

Taphrospilus 1 species | *p. 268*
Many-spotted Hummingbird
T. hypostictus | Ecuador

Leucochloris 1 species | *p. 268*
White-throated Hummingbird
L. albicollis | Brazil

Leucippus 4 species | *p. 268*
Tumbes Hummingbird *L. baeri* | Peru

Amazilia 44 species | *p. 268*
Rufous-throated Hummingbird
A. sapphirina | Brazil

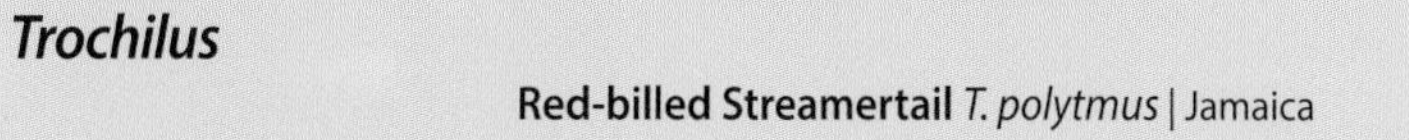

Trochilus 2 species | *p. 270*
Red-billed Streamertail *T. polytmus* | Jamaica

Subfamily **Trochilinae** Emeralds ; Mountain-gems ; Bees

Goethalsia 1 species | *p.270*
Pirre Hummingbird
G. bella | Panama

Goldmania 1 species | *p.270*
Violet-capped Hummingbird
G. violiceps | Panama

Basilinna 2 species | *p.270*
White-eared Hummingbird
B. leucotis | Guatemala

Hylonympha 1 species | *p.271*
Scissor-tailed Hummingbird
H. macrocerca | Venezuela

Sternoclyta 1 species | *p.271*
Violet-chested Hummingbird *S. cyanopectus* | Venezuela

Eugenes 1 species | *p.271*
Magnificent Hummingbird
E. fulgens | Costa Rica

Panterpe 1 species | *p.271*
Fiery-throated Hummingbird
P. insignis | Costa Rica

3/4

Heliomaster 4 species | *p. 271*
Long-billed Starthroat
H. longirostris | Colombia

Lampornis 8 species | *p. 271*
White-throated Mountain-gem
L. castaneoventris | Panama

Lamprolaima 1 species | *p. 271*
Garnet-throated Hummingbird
L. rhami | Guatemala

Myrtis 1 species | *p. 272*
Purple-collared Woodstar *M. fanny* | Peru

Eulidia 1 species | *p. 272*
Chilean Woodstar *E. yarrellii* | Chile

Rhodopis 1 species | *p. 272*
Oasis Hummingbird *R. vesper* | Chile

Thaumastura 1 species | *p. 272*
Peruvian Sheartail *T. cora* | Peru

Subfamily **Trochilinae** Emeralds ; Mountain-gems ; Bees

Chaetocercus 6 species | *p. 272*
Gorgeted Woodstar
C. heliodor | Colombia

Myrmia 1 species | *p. 272*
Short-tailed Woodstar
M. micrura | Peru

Microstilbon 1 species | *p. 272*
Slender-tailed Woodstar
M. burmeisteri | Argentina

Doricha 2 species | *p. 273*
Slender Sheartail *D. enicura* | Guatemala

Calliphlox 3 species | *p. 272*
Purple-throated Woodstar *C. mitchellii* | Ecuador

Tilmatura 1 species | *p. 273*
Sparkling-tailed Woodstar *T. dupontii* | Guatemala

Calothorax 2 species | *p. 273*
Lucifer Hummingbird *C. lucifer* | USA

4/4

Nesophlox 1 species | *p.273*
Bahama Hummingbird *N. evelynae* | Bahamas

Mellisuga 2 species | *p.273*
Bee Hummingbird *M. helenae* | Cuba

Calypte 2 species | *p.273*
Anna's Hummingbird *C. anna* | Canada

Archilochus 2 species | *p.273*
Ruby-throated Hummingbird *A. colubris* | USA

Selasphorus 7 species | *p.273*
Scintillant Hummingbird *S. scintilla* | Costa Rica

Atthis 2 species | *p.273*
Bumblebee Hummingbird *A. heloisa* | Mexico

Coding used in the complete list of hummingbird species that follows:

Genus (pale green boxes)**:** one representative of each genus is shown on the *preceding pages*; the position of the relevant image on the page is indicated as: T = top, M = middle row, B = bottom, L = left, C = center, R = right.

IUCN Red List category (coded square boxes)**:** see *page 192* for details of the codes used for the various threat categories.

Population trend: increasing ↑, stable **=**, decreasing ↓, unknown **?**

Species distribution: broad indication of main distribution, qualified with compass direction/general area, as appropriate: N = north, S = south, E = east, W = west, C = central; for species that breed only in one country (country endemics), the name of the country is highlighted in **bold**.

Maps: resident ■, winter ■, breeding ■, on migration ■.

Taxonomic notes:
A number of the species that BirdLife International recognizes are not included on other taxonomic lists, and *vice versa*. For the sake of completeness, those used throughout this book that differ from the latest *IOC World Bird List* (IOC), *eBird/Clements* checklist (eBird) or *Howard and Moore* checklist (H&M) are highlighted, with footnotes. Other taxonomic lists use different scientific names for certain species, but covering these is beyond the scope of this present book: a useful comparison of taxonomies and nomenclature can, however, be downloaded from https://www.worldbirdnames.org.

Black Jacobin *Florisuga fusca* | Brazil

Subfamily **Florisuginae**

Topaza — 2 species, *p. 242* TL

LC ↓ **Crimson Topaz** *T. pella*
3 sspp. 0–600 m
SE Venezuela to N Brazil; W Brazil
pp. 72, 225, 242

LC ↓ **Fiery Topaz** *T. pyra*
2 sspp. 0–500 m
Amazonia
NOT ILLUSTRATED
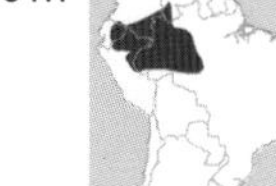

Florisuga — 2 species, *p. 242* ML

LC ↓ **White-necked Jacobin** *F. mellivora*
2 sspp. 0–1,900 m
SE Mexico to NW Brazil/Bolivia
pp. 13, 79, 154, 242

LC ? **Black Jacobin** *F. fusca*
Monotypic 0–1,400 m
SE Brazil to Uruguay; SE Paraguay to NE Argentina
pp. 24, 79, 169, 256

Subfamily **Phaethornithinae**

Eutoxerxes — 2 species, *p. 242* TC

LC ? **White-tipped Sicklebill** *E. aquila*
3 sspp. 300–2,300 m
Costa Rica to N Peru
pp. 40, 235, 242

LC ↓ **Buff-tailed Sicklebill** *E. condamini*
2 sspp. 180–2,800 m
SW Colombia to W Bolivia
NOT ILLUSTRATED

Ramphodon — 1 species, *p. 242* TR

LC ↓ **Saw-billed Hermit** *R. naevius*
Monotypic 0–900 m
SE **Brazil**
pp. 45, 173, 242

Glaucis — 3 species, *p. 242* MC

VU ↓ **Hook-billed Hermit** *G. dohrnii*
Monotypic 0–500 m
E **Brazil**
pp. 103, 205

LC ↓ **Bronzy Hermit** *G. aeneus*
Monotypic 0–800 m
NE Honduras to NW Ecuador
NOT ILLUSTRATED
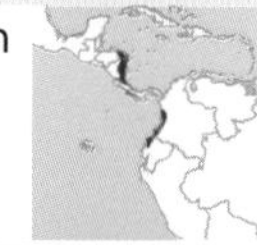

LC ↓ **Rufous-breasted Hermit** *G. hirsutus*
2 sspp. 0–1,800 m
Panama to Amazonia; E Brazil
p. 242

Threnetes — 3 species, *p. 242* MR

LC ↓ **Band-tailed Barbthroat** *T. ruckeri*
3 sspp. 0–1,200 m
Belize to W Venezuela/ Ecuador
p. 242

LC = **Pale-tailed Barbthroat** *T. leucurus*
5 sspp. 0–1,600 m
E Colombia to N Brazil/ N Bolivia
p. 165

LC ? **Sooty Barbthroat** *T. niger*
Monotypic 0–500 m
French Guiana/N Brazil
NOT ILLUSTRATED

Anopetia — 1 species, *p. 242* BC

LC ? **Broad-tipped Hermit** *A. gounellei*
Monotypic 40–1,200 m
NE **Brazil**
p. 242

Phaethornis — 30 species, *p. 242* BR

LC ? **Dusky-throated Hermit** *P. squalidus*
Monotypic 0–2,300 m
SE **Brazil**
p. 172

LC ? **Streak-throated Hermit** *P. rupurumii*
2 sspp. 0–500 m
S Venezuela to Guyana/ N Brazil
p. 89

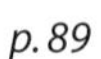

LC = **Little Hermit** *P. longuemareus*
Monotypic 0–700 m
NE Venezuela/Trinidad to French Guiana
p. 163

VU ↓ **Tapajos Hermit** *P. aethopygus*
Monotypic 0–500 m
NC **Brazil**
NOT ILLUSTRATED

LC ↓ **Minute Hermit** *P. idaliae*
Monotypic 0–500 m
E **Brazil**
p. 172

LC ↓ **Cinnamon-throated Hermit** *P. nattereri*
Monotypic 0–500 m
NE Brazil; W Brazil/E Bolivia
p. 274

LC ↓ **Black-throated Hermit** *P. atrimentalis*
2 sspp. 0–1,200 m
C Colombia to C Peru
NOT ILLUSTRATED

LC ↓ **Stripe-throated Hermit** *P. striigularis*

4 sspp. 0–1,800 m
SE Mexico to N Venezuela/Ecuador

NOT ILLUSTRATED

LC ? **Grey-chinned Hermit** *P. griseogularis*

2 sspp. 300–2,100 m
W Venezuela to N Peru; S Venezuela/NW Brazil

p. 13

LC ? **Porculla Hermit** [1] *P. porcullae*

Monotypic 400–2,100 m
S Ecuador to NW Peru

NOT ILLUSTRATED

LC ↓ **Reddish Hermit** *P. ruber*

4 sspp. 0–1,500 m
E Venezuela to N Bolivia/Brazil

pp. 41, 89, 103

LC ↓ **White-browed Hermit** *P. stuarti*

Monotypic 300–1,600 m
C Peru to C Bolivia

NOT ILLUSTRATED

LC ↓ **Buff-bellied Hermit** *P. subochraceus*

Monotypic 80–800 m
C Bolivia to W Brazil

NOT ILLUSTRATED

LC = **Sooty-capped Hermit** *P. augusti*

3 sspp. 200–2,500 m
NE Colombia to N Venezuela; E Venez./W Guyana/N Brazil

NOT ILLUSTRATED

LC ? **Planalto Hermit** *P. pretrei*

Monotypic 400–2,100 m
E Bolivia/Paraguay/N Argentina to NE Brazil

p. 257

LC ↓ **Scale-throated Hermit** *P. eurynome*

2 sspp. 100–2,250 m
E Paraguay/NE Argentina to SE Brazil

p. 242

LC ↓ **Pale-bellied Hermit** *P. anthophilus*

2 sspp. 0–1,200 m
E Panama to Colombia/N Venezuela

NOT ILLUSTRATED

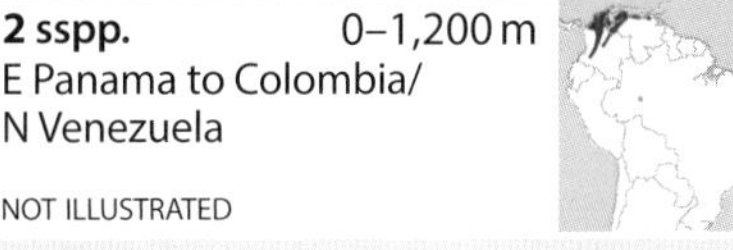

LC ↓ **White-bearded Hermit** *P. hispidus*

Monotypic 0–1,200 m
W Amazonia

NOT ILLUSTRATED

LC ? **White-whiskered Hermit** *P. yaruqui*

Monotypic 0–2,000 m
E Panama to Ecuador

p. 155

LC ↓ **Green Hermit** *P. guy*

4 sspp. 300–2,200 m
Costa Rica to W Venezuela/Peru; NE Venezuela/Trinidad

pp. 74, 221

LC ? **Tawny-bellied Hermit** *P. syrmatophorus*

2 sspp. 900–2,500 m
WC Colombia to N Peru

NOT ILLUSTRATED

NT ↓ **Koepcke's Hermit** *P. koepckeae*

Monotypic 450–1,300 m
Peru

p. 155

LC ? **Needle-billed Hermit** *P. philippii*

Monotypic 0–500 m
E Peru to W Brazil/NW Bolivia

NOT ILLUSTRATED

LC = **Straight-billed Hermit** *P. bourcieri*

Monotypic 0–1,600 m
E Peru to N Brazil/French Guiana

NOT ILLUSTRATED

LC = **Ash-bellied Hermit** [2] *P. major*

Monotypic 0–400 m
NC **Brazil**

NOT ILLUSTRATED

LC ↓ **Mexican Hermit** [3] *P. mexicanus*

2 sspp. 0–2,000 m
W & S **Mexico**

NOT ILLUSTRATED

LC ? **Long-billed Hermit** *P. longirostris*

3 sspp. 0–2,500 m
S Mexico to NW Venezuela

pp. 33, 131

LC ? **Ecuadorian Hermit** [4] *P. baroni*

Monotypic 0–1,300 m
W Ecuador to NW Peru

p. 9

LC ↓ **Long-tailed Hermit** *P. superciliosus*

2 sspp. 0–2,500 m
S Venezuela to N Brazil

p. 61

LC ↓ **Great-billed Hermit** *P. malaris*

6 sspp. 0–1,800 m
W Amazonia; Suriname/N Brazil; E Brazil

pp. 89, 102

Planalto Hermit *Phaethornis pretrei* | Brazil

[1] **Porculla Hermit** *Phaethornis porcullae* is not recognized by IOC, eBird or H&M, all of which treat it as a subspecies of Grey-chinned Hermit *P. griseogularis*.

[2] **Ash-bellied Hermit** *Phaethornis major* is not recognized by IOC, eBird or H&M, all of which treat it as a subspecies of Straight-billed Hermit *P. bourcieri*.

[3] **Mexican Hermit** *Phaethornis mexicanus* is not recognized by H&M, which treats it as a subspecies of Long-billed Hermit *P. longirostris*.

[4] **Ecuadorian Hermit** *Phaethornis baroni* is not recognized by IOC, eBird or H&M, all of which treat it as a subspecies of Long-billed Hermit *P. longirostris*.

Subfamily **Polytminae**

Doryfera — 2 species, p. 243 TL

LC ? Green-fronted Lancebill *D. ludovicae*

2 sspp. 750–2,600 m
Costa Rica to W Bolivia/ W Venezuela

pp. 151, 243

LC ↓ Blue-fronted Lancebill *D. johannae*

2 sspp. 300–1,800 m
C Colombia to C Peru; S Venezuela to W Guyana

NOT ILLUSTRATED

Schistes [5] — 2 species, p. 243 TC

LC = Western Wedge-billed Hummingbird *S. albogularis*

Monotypic 800–2,600 m
W Colombia to Ecuador

pp. 39, 137, 243

LC = Eastern Wedge-billed Hummingbird *S. geoffroyi*

2 sspp. 500–2,600 m
W Venezuela to W Bolivia

p. 137

Augastes — 2 species, p. 243 TR

LC ↓ Hyacinth Visorbearer *A. scutatus*

3 sspp. 900–2,000 m
EC **Brazil**

pp. 167, 243

NT ↓ Hooded Visorbearer *A. lumachella*

Monotypic 750–2,000 m
EC **Brazil**

pp. 96, 167

Brown Violet-ear *Colibri delphinae* | Ecuador

Colibri — 4 species, p. 243 ML

LC ↓ Brown Violet-ear *C. delphinae*

Monotypic 100–2,800 m
Belize to Panama; N Venez./Trinidad to Bolivia; S Venez./N Brazil/W Suriname; E Brazil

pp. 95, 146, 258

LC ↓ Green Violet-ear [6] *C. thalassinus*

5 sspp. 600–3,250 m
C Mexico to NW Nicaragua; Costa Rica to C Panama; NE Venezuela to C Bolivia

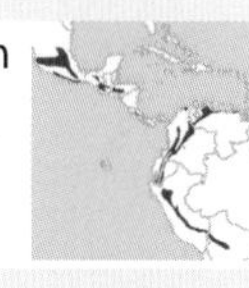

pp. 67, 75

LC ? Sparkling Violet-ear *C. coruscans*

2 sspp. 1,700–4,500 m
NW Venezuela to N Chile/ NW Argentina; S Venezuela/ E Guyana/N Brazil

pp. 99, 111, 141

LC ? White-vented Violet-ear *C. serrirostris*

Monotypic 0–2,900 m
Bolivia/NW Argentina to S Brazil/NE Paraguay

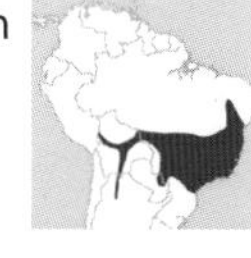

p. 243

Androdon — 1 species, p. 243 MC

LC ↓ Tooth-billed Hummingbird *A. aequatorialis*

Monotypic 300–1,100 m
E Panama to NW Ecuador

pp. 45, 243

Heliactin — 1 species, p. 243 MR

LC ↑ Horned Sungem *H. bilophus*

Monotypic 0–1,000 m
N Bolivia to NE & SE Brazil

pp. 13, 110, 167, 243

Heliothryx — 2 species, p. 243 BL

LC ↓ Purple-crowned Fairy *H. barroti*

Monotypic 0–1,100 m
SE Mexico to Ecuador

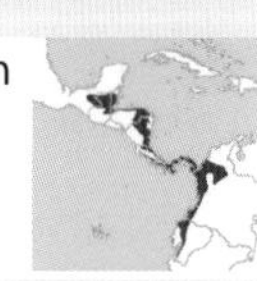

pp. 87, 220, 243

LC ↓ Black-eared Fairy *H. auritus*

3 sspp. 0–1,100 m
Amazonia to French Guiana; SE Brazil

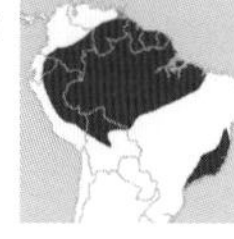

NOT ILLUSTRATED

Polytmus — 3 species, p. 243 BC

LC ? White-tailed Goldenthroat *P. guainumbi*

3 sspp. 0–600 m
C Colombia to NE Brazil to NE Argentina/Paraguay to E Peru

p. 243

LC ? Tepui Goldenthroat *P. milleri*

Monotypic 1,300–2,200 m
SE Venezuela to W Guyana/ N Brazil

NOT ILLUSTRATED

LC ? Green-tailed Goldenthroat *P. theresiae*

2 sspp. 0–300 m
E Colombia/S Venezuela to French Guiana/N Brazil

pp. 111, 163

Chrysolampis — 1 species, p. 243 BR

LC ? Ruby-topaz Hummingbird *C. mosquitus*

Monotypic 0–1,700 m
E Panama to Brazil to E Bolivia

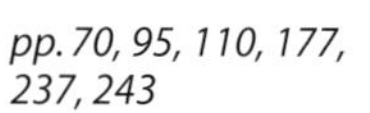

pp. 70, 95, 110, 177, 237, 243

Avocettula — 1 species, p. 244 TL

LC ↓ Fiery-tailed Awlbill *A. recurvirostris*

Monotypic 0–1,200 m
E Ecuador; NE Venezuela to N Brazil

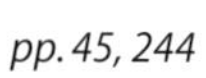

pp. 45, 244

Anthracothorax — 8 species, p. 244 ML

LC ↓ Green-throated Mango *A. viridigula*

Monotypic 0–500 m
NE Venezuela to N Brazil

NOT ILLUSTRATED

LC ↓ Green-breasted Mango *A. prevostii*

4 sspp. 0–1,200 m
S & EC Mexico to NW Panama; NE Colombia to N Venezuela

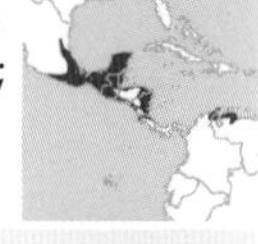

pp. 21, 121, 130

LC = Black-throated Mango *A. nigricollis*

2 sspp. 0–1,900 m
Panama to NE Argentina/ N Uruguay

pp. 161, 244

[5] The two species of *Schistes* are treated as one by H&M, named Wedge-billed Hummingbird *S. geoffroyi*, with three subspecies.

[6] **Green Violet-ear** *Colibri thalassinus* is split into two species by IOC and eBird: Mexican Violetear/Violet-ear *C. thalassinus* (C Mexico to NW Nicaragua) and Lesser Violetear/Violet-ear *C. cyanotus* (Costa Rica to C Panama; NE Venezuela to C Bolivia), with 4 sspp.

[7] **Puerto Rican Mango** *Anthracothorax aurulentus* is not recognized by IOC, eBird or H&M, all of which treat it as a subspecies of what BirdLife call Hispaniolan Mango *A. dominicus*, but use the name Antillean Mango.

LC = **Veraguas Mango** *A. veraguensis*

Monotypic 0–500 m
SE Costa Rica to S Panama

NOT ILLUSTRATED

LC ? **Hispaniolan Mango** *A. dominicus*

Monotypic 0–2,600 m
Haiti/Dominican Republic (Hispaniola)

p. 127

LC ↓ **Puerto Rican Mango** [7] *A. aurulentus*

Monotypic 0–800 m
Puerto Rico/British Virgin Islands/Virgin Islands

NOT ILLUSTRATED

LC ? **Green Mango** *A. viridis*

Monotypic 500–1,200 m
Puerto Rico

NOT ILLUSTRATED

LC ? **Jamaican Mango** *A. mango*

Monotypic 0–1,500 m
Jamaica

pp. 124, 259

Eulampis

2 species
p. 244 BL

LC ? **Green-throated Carib** *E. holosericeus*

2 sspp. 0–1,000 m
Puerto Rico to Grenada/Barbados

pp. 82, 126

LC ? **Purple-throated Carib** *E. jugularis*

Monotypic 800–1,200 m
St Barthélemy to Grenada

pp. 85, 126, 244

Subfamily **Lesbiinae**

Heliangelus

9 species
p. 244 TC

LC = **Orange-throated Sunangel** *H. mavors*

Monotypic 2,000–3,200 m
W Venezuela to NC Colombia

p. 70

LC ? **Merida Sunangel** [8] *H. spencei*

Monotypic 2,000–3,600 m
W **Venezuela**

NOT ILLUSTRATED

LC = **Longuemare's Sunangel** [9] *H. clarisse*

3 sspp. 1,800–3,300 m
NE Colombia/NW Venezuela; W Venezuela to C Colombia

NOT ILLUSTRATED

LC = **Amethyst-throated Sunangel** *H. amethysticollis*

4 sspp. 1,950–3,700 m
S Ecuador to C Bolivia

pp. 50, 244

LC ↓ **Gorgeted Sunangel** *H. strophianus*

Monotypic 1,200–2,800 m
SW Colombia to NW Ecuador; SW Ecuador

p. 74

LC = **Tourmaline Sunangel** *H. exortis*

Monotypic 1,500–3,400 m
WC Colombia to C Ecuador

pp. 42, 105, 152

LC ? **Little Sunangel** *H. micraster*

2 sspp. 2,200–3,100 m
S Ecuador to NW Peru

p. 259

LC = **Purple-throated Sunangel** *H. viola*

Monotypic 1,800–3,300 m
C Ecuador to NW Peru

pp. 54, 217

EN ↓ **Royal Sunangel** *H. regalis*

2 sspp. 550–2,200 m
SE Ecuador to NW Peru

pp. 74, 203

Jamaican Mango *Anthracothorax mango* | Jamaica

Little Sunangel *Heliangelus micraster* | Ecuador

Sephanoides

2 species
p. 244 TR

LC ↓ **Green-backed Firecrown** *S. sephaniodes*

Monotypic 0–2,000 m
S Chile/S Argentina

pp. 158, 244

CR ↓ **Juan Fernandez Firecrown** *S. fernandensis*

Monotypic 100–800 m
Chile (Juan Fernández Is.)

p. 204

[8] **Merida Sunangel** *Heliangelus spencei* not recognized by eBird, which treats it as a is subspecies of Amethyst-throated Sunangel *H. amethysticollis*.

[9] **Longuemare's Sunangel** *Heliangelus clarisse* is not recognized by eBird or H&M, both of which treat it as a subspecies of Amethyst-throated Sunangel *H. amethysticollis*.

Discosura

5 species
p. 244 MC

LC ↓ **Green Thorntail** *D. conversii*

Monotypic 0–1,500 m
Costa Rica to SW Ecuador

pp. 47, 149, 244

NT **?** **Wire-crested Thorntail** *D. popelairii*

Monotypic 400–1,500 m
C Colombia; SW Colombia to W Bolivia

pp. 93, 202

LC ↓ **Black-bellied Thorntail** *D. langsdorffi*

2 sspp. 100–800 m
W Amazonia; SE Brazil

NOT ILLUSTRATED

DD **?** **Coppery Thorntail** *D. letitiae*

Monotypic ? m
Bolivia?

NOT ILLUSTRATED

LC ↓ **Racket-tailed Coquette** *D. longicaudus*

Monotypic 0–900 m
E Colombia to French Guiana/N Brazil; E Brazil

p. 96

Lophornis

11 species
p. 244 MR

LC = **Tufted Coquette** *L. ornatus*

Monotypic 100–1,000 m
C Venezuela to N Brazil

pp. 6, 10, 91

Rufous-crested Coquette *Lophornis delattrei* | Peru

NT ↓ **Dot-eared Coquette** *L. gouldii*

Monotypic 0–500 m
NC **Brazil** to E Bolivia (?)

NOT ILLUSTRATED

LC **?** **Frilled Coquette** *L. magnificus*

Monotypic 0–1,100 m
S **Brazil**

pp. 169, 236

CR ↓ **Short-crested Coquette** *L. brachylophus*

Monotypic 900–1,800 m
S **Mexico**

p. 196

LC ↓ **Rufous-crested Coquette** *L. delattrei*

2 sspp. 600–2,000 m
Panama to C Bolivia

pp. 61, 97, 260

LC **?** **Spangled Coquette** *L. stictolophus*

Monotypic 100–1,300 m
W Venezuela to NW Peru

p. 244

NT **?** **Festive Coquette** *L. chalybeus*

Monotypic 0–1,000 m
SE **Brazil**

pp. 29, 76, 171

LC ↓ **Butterfly Coquette** [10] *L. verreauxii*

2 sspp. 0–1,000 m
W Amazonia to C Venezuela

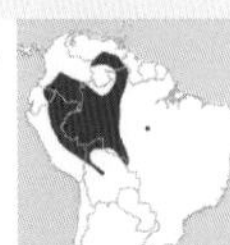

NOT ILLUSTRATED

LC **?** **Peacock Coquette** *L. pavoninus*

2 sspp. 500–2,000 m
W Amazonia to C Venezuela

NOT ILLUSTRATED

LC **?** **Black-crested Coquette** *L. helenae*

Monotypic 100–1,450 m
S Mexico to Costa Rica

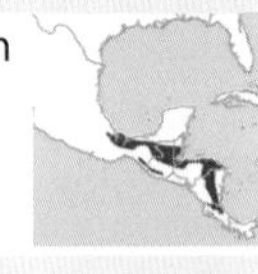

pp. 83, 91, 129

LC ↓ **White-crested Coquette** *L. adorabilis*

Monotypic 300–1,200 m
S Costa Rica to SW Panama

p. 41

Phlogophilus

2 species
p. 244 BC

VU ↓ **Ecuadorian Piedtail** *P. hemileucurus*

Monotypic 600–1,500 m
SW Colombia to NW Peru

pp. 202, 244

NT ↓ **Peruvian Piedtail** *P. harterti*

Monotypic 750–1,500 m
CS **Peru**

NOT ILLUSTRATED

Adelomyia

1 species
p. 244 BR

LC **?** **Speckled Hummingbird** *A. melanogenys*

8 sspp. 1,000–2,500 m
N Venezuela to NW Argentina

pp. 104, 151, 244

Aglaiocercus

3 species
p. 245 TL

LC ↓ **Long-tailed Sylph** *A. kingii*

6 sspp. 1,200–3,000 m
W Venezuela to C Bolivia

pp. 18, 61, 113, 228, 245

LC = **Violet-tailed Sylph** *A. coelestis*

2 sspp. 900–2,100 m
WC Colombia to C Ecuador

pp. 86, 147

EN ↓ **Venezuelan Sylph** *A. berlepschi*

Monotypic 1,450–1,800 m
NE **Venezuela**

NOT ILLUSTRATED

Sappho

1 species
p. 245 TC

LC = **Red-tailed Comet** *S. sparganurus*

2 sspp. 1,500–4,000 m
W Bolivia to N Argentina

pp. 48, 159, 245

Taphrolesbia

1 species
p. 245 TR

EN ↓ **Grey-bellied Comet** *T. griseiventris*

Monotypic 2,750–3,500 m
NW **Peru**

pp. 203, 245

Polyonymus
1 species p. 245 MR

LC = **Bronze-tailed Comet** *P. caroli*

Monotypic 1,500–3,600 m
W **Peru**
p. 245

Oreotrochilus
7 species p. 245 BL

LC = **Ecuadorian Hillstar** *O. chimborazo*

3 sspp. 3,500–5,200 m
SW Colombia to Ecuador
pp. 55, 218, 245

LC = **Andean Hillstar** *O. estella*

2 sspp. 2,400–5,000 m
S Peru to NW Argentina
p. 108

LC = **Green-headed Hillstar** [11] *O. stolzmanni*

Monotypic 3,600–4,200 m
S Ecuador to WC Peru
NOT ILLUSTRATED

CR ↓ **Blue-throated Hillstar** [12] *O. cyanolaemus*

Monotypic 3,300–3,700 m
CS **Ecuador**
pp. 4, 207

LC = **White-sided Hillstar** *O. leucopleurus*

Monotypic 1,200–4,050 m
S Bolivia to SW Argentina/ SE Chile
p. 158

LC = **Black-breasted Hillstar** *O. melanogaster*

Monotypic 3,500–4,200 m
SW **Peru**
p. 207

LC ↓ **Wedge-tailed Hillstar** *O. adela*

Monotypic 2,100–4,000 m
WC Bolivia to NW Argentina
p. 205

Opisthoprora
1 species p. 245 BR

LC = **Mountain Avocetbill** *O. euryptera*

Monotypic 2,600–3,600 m
SW Colombia to WC Peru
p. 245

Lesbia
2 species p. 246 TL

LC = **Black-tailed Trainbearer** *L. victoriae*

3 sspp. 2,600–4,000 m
SW Colombia to SE Peru
p. 14

LC = **Green-tailed Trainbearer** *L. nuna*

7 sspp. 1,700–3,800 m
C Colombia; SW Colombia to W Bolivia
p. 246

Ramphomicron
2 species p. 246 TR

EN ↓ **Black-backed Thornbill** *R. dorsale*

Monotypic 2,000–4,600 m
N **Colombia**
pp. 71, 138, 194

LC ↓ **Purple-backed Thornbill** *R. microrhynchum*

4 sspp. 1,700–3,400 m
N Colombia; W Venezuela to NW Peru; SW Peru; W Bolivia
pp. 41, 246

Chalcostigma
5 species p. 246 ML

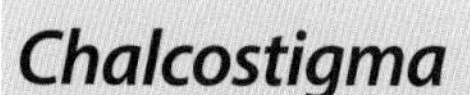

LC = **Rufous-capped Thornbill** *C. ruficeps*

Monotypic 1,800–3,300 m
S Ecuador to C Bolivia
p. 246

LC = **Olivaceous Thornbill** *C. olivaceum*

2 sspp. 3,600–4,600 m
C Peru to W Bolivia
p. 140

LC ↓ **Blue-mantled Thornbill** *C. stanleyi*

3 sspp. 2,800–4,200 m
N Ecuador; NW Peru to W Bolivia
p. 261

LC ↓ **Bronze-tailed Thornbill** *C. heteropogon*

Monotypic 3,000–3,900 m
W Venezuela to C Colombia
NOT ILLUSTRATED

LC ↓ **Rainbow-bearded Thornbill** *C. herrani*

2 sspp. 2,700–4,000 m
WC Colombia to NW Peru
pp. 81, 101

Blue-mantled Thornbill *Chalcostigma stanleyi* | Ecuador

Oxypogon [13]
4 species p. 246 MC

CR ↓ **Blue-bearded Helmetcrest** *O. cyanolaemus*

Monotypic 3,000–4,800 m
N **Colombia**
NOT ILLUSTRATED

LC ↓ **White-bearded Helmetcrest** *O. lindenii*

Monotypic 2,900–4,600 m
W **Venezuela**
NOT ILLUSTRATED

LC ↓ **Green-bearded Helmetcrest** *O. guerinii*

Monotypic 3,000–5,200 m
NC **Colombia**
pp. 140, 246

VU ↓ **Buffy Helmetcrest** *O. stuebelii*

Monotypic 3,300–4,800 m
WC **Colombia**
p. 224

[10] **Butterfly Coquette** *Lophornis verreauxii* is not recognized by eBird or H&M, both of which treat it as a subspecies of Festive Coquette *L. chalybeus*.

[11] **Green-headed Hillstar** *Oreotrochilus stolzmanni* is not recognized by H&M, which treats it as a subspecies of Andean Hillstar *O. estella*.

[12] **Blue-throated Hillstar** *Oreotrochilus cyanolaemus* is not (yet) recognized by H&M (newly described species).

[13] The four species of ***Oxypogon*** are treated as one by H&M, named Bearded Helmetcrest *O. guerinii*, with 4 sspp.

Scaled Metaltail *Metallura aeneocauda* | Peru

Oreonympha [14]

2 species
p. 246 MR

LC = **Western Mountaineer** *O. albolimbata*

Monotypic 2,500–3,900 m
WCS **Peru**

NOT ILLUSTRATED

LC = **Eastern Mountaineer** *O. nobilis*

Monotypic 2,500–3,900 m
ECS **Peru**

pp. 141, 246

Metallura

9 species
p. 246 BL

EN ↓ **Perija Metaltail** *M. iracunda*

Monotypic 2,400–3,200 m
NE Colombia/NW Venezuela

NOT ILLUSTRATED

LC = **Tyrian Metaltail** *M. tyrianthina*

7 sspp. 1,500–4,200 m
N Colombia; N Venezuela to C Bolivia

pp. 113, 142

LC ↓ **Viridian Metaltail** *M. williami*

4 sspp. 2,700–3,800 m
C Colombia to S Ecuador

p. 246

EN ↓ **Violet-throated Metaltail** *M. baroni*

Monotypic 3,100–4,000 m
CS **Ecuador**

p. 201

LC ↓ **Neblina Metaltail** *M. odomae*

Monotypic 2,850–3,350 m
S Ecuador to NW Peru

NOT ILLUSTRATED

LC ↓ **Coppery Metaltail** *M. theresiae*

2 sspp. 2,900–3,800 m
NW **Peru**

p. 53

LC ↓ **Fire-throated Metaltail** *M. eupogon*

Monotypic 2,900–4,000 m
CS **Peru**

NOT ILLUSTRATED

LC ↓ **Scaled Metaltail** *M. aeneocauda*

2 sspp. 2,500–3,600 m
SE Peru to C Bolivia

p. 262

LC ↓ **Black Metaltail** *M. phoebe*

Monotypic 1,500–4,500 m
W **Peru**

p. 28

Haplophaedia

3 species
p. 246 BC

LC = **Greenish Puffleg** *H. aureliae*

6 sspp. 1,500–3,100 m
E Panama to NW Peru

p. 246

LC ↓ **Buff-thighed Puffleg** *H. assimilis*

2 sspp. 1,500–2,600 m
NW Peru to W Bolivia

NOT ILLUSTRATED

Black-thighed Puffleg *Eriocnemis derbyi* | Colombia

NT ↓ **Hoary Puffleg** *H. lugens*

Monotypic 1,200–2,000 m
SW Colombia to N Ecuador

p. 148

Eriocnemis

12 species
p. 246 BR

EN ↓ **Black-breasted Puffleg** *E. nigrivestis*

Monotypic 2,700–3,500 m
N **Ecuador**

pp. 201, 231

CR ↓ **Gorgeted Puffleg** *E. isabellae*

Monotypic 2,600–3,200 m
SW **Colombia**

NOT ILLUSTRATED

LC = **Glowing Puffleg** *E. vestita*

4 sspp. 2,200–3,700 m
W Venezuela to NW Peru

p. 76

NT ↓ **Black-thighed Puffleg** *E. derbyi*

Monotypic 2,500–3,600 m
C Colombia to N Ecuador

p. 262

CR ? **Turquoise-throated Puffleg** *E. godini*

Monotypic 2,100–2,300 m
N **Ecuador** (possibly Extinct)

NOT ILLUSTRATED

NT ↓ **Coppery-bellied Puffleg** *E. cupreoventris*

Monotypic 1,900–3,200 m
W Venezuela to C Colombia

p. 52

LC = **Sapphire-vented Puffleg** *E. luciani*

3 sspp. 2,800–4,800 m
W Venezuela?; SW Colombia to C Ecuador

p. 219

LC = **Coppery-naped Puffleg** [15] *E. sapphiropygia*

2 sspp. 2,400–3,500 m
Peru

NOT ILLUSTRATED

LC = **Golden-breasted Puffleg** *E. mosquera*

Monotypic 1,200–3,600 m
WC Colombia; SW Colombia to C Ecuador

pp. 52, 246

LC = **Blue-capped Puffleg** *E. glaucopoides*

Monotypic 1,500–3,400 m
C Bolivia to N Argentina

NOT ILLUSTRATED

EN ↓ **Colorful Puffleg** *E. mirabilis*

Monotypic 2,200–2,800 m
SW **Colombia**

NOT ILLUSTRATED

LC ↓ **Emerald-bellied Puffleg** *E. aline*

2 sspp. 2,300–2,800 m
C Colombia to C Peru

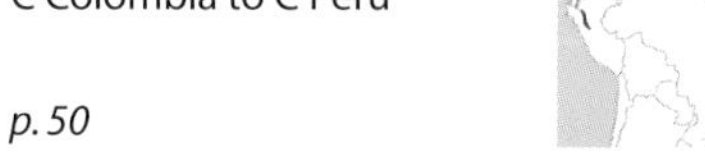

p. 50

Loddigesia

1 species
p. 247 TL

EN ↓ **Marvelous Spatuletail** *L. mirabilis*

Monotypic 2,100–2,900 m
NW **Peru**

pp. 27, 208, 223, 247

Aglaeactis

4 species
p. 247 TR

LC = **Shining Sunbeam** *A. cupripennis*

2 sspp. 2,500–4,300 m
Colombia to Peru

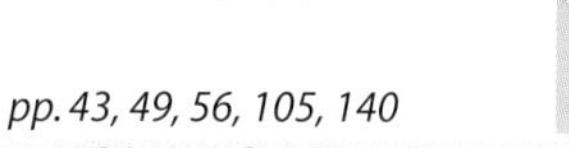

pp. 43, 49, 56, 105, 140

NT ↓ **White-tufted Sunbeam** *A. castelnaudii*

2 sspp. 3,500–4,300 m
C & CS **Peru**

p. 247

VU ↓ **Purple-backed Sunbeam** *A. aliciae*

Monotypic 2,900–3,500 m
WC **Peru**

NOT ILLUSTRATED

LC = **Black-hooded Sunbeam** *A. pamela*

Monotypic 2,900–3,900 m
CW **Bolivia**

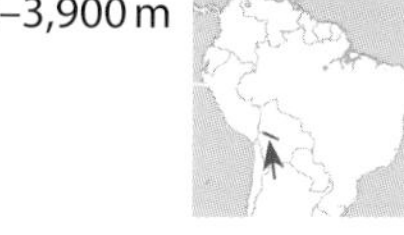

p. 224

Coeligena

19 species
p. 247 ML

LC = **Bronzy Inca** *C. coeligena*

6 sspp. 1,500–2,600 m
N Venezuela to C Bolivia

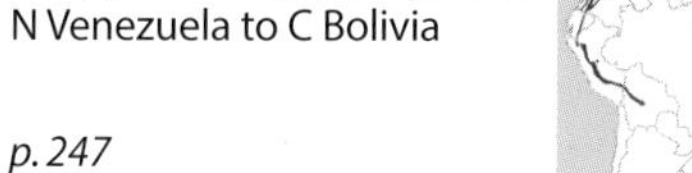

p. 247

VU ↓ **Black Inca** *C. prunellei*

Monotypic 1,000–2,800 m
C **Colombia**

p. 200

LC ↓ **Brown Inca** *C. wilsoni*

Monotypic 700–2,200 m
WC Colombia to SW Ecuador

pp. 32, 146

LC ↓ **Green Inca** [16] *C. conradii*

Monotypic 1,500–3,000 m
W Venezuela to NE Colombia

NOT ILLUSTRATED

LC ↓ **Collared Inca** *C. torquata*

4 sspp. 1,500–3,000 m
W Venezuela to S Peru

pp. 32, 59

LC ↓ **Vilcabamba Inca** [17] *C. eisenmanni*

Monotypic 2,070–2,840 m
CS **Peru**

NOT ILLUSTRATED

LC ↓ **Gould's Inca** [18] *C. inca*

2 sspp. 1,600–3,200 m
SE Peru to W Bolivia

NOT ILLUSTRATED

LC ↓ **Huanuco Starfrontlet** [19] *C. dichroura*

Monotypic 1,900–3,700 m
NW Peru to W Bolivia

NOT ILLUSTRATED

Cuzco Starfrontlet *Coeligena osculans* | Peru

LC ↓ **Apurimac Starfrontlet** [20] *C. albicaudata*

Monotypic 2,250–3,600 m
CS **Peru**

NOT ILLUSTRATED

LC ↓ **Cuzco Starfrontlet** [21] *C. osculans*

Monotypic 2,000–3,700 m
SE Peru to W Bolivia

p. 263

LC ↓ **Bolivian Starfrontlet** *C. violifer*

Monotypic 1,300–3,700 m
CW **Bolivia**

NOT ILLUSTRATED

LC ↓ **Rainbow Starfrontlet** *C. iris*

6 sspp. 1,700–4,000 m
S Ecuador to WC Peru

pp. 54, 217

[14] The two ***Oreonympha*** species are treated as one by IOC, eBird and H&M, named Bearded Mountaineer *O. nobilis*, with 2 sspp.

[15] **Coppery-naped Puffleg** *Eriocnemis sapphiropygia* is not recognized by IOC, eBird or H&M, all of which treat it as a subspecies of Sapphire-vented Puffleg *E. luciani*.

[16] **Green Inca** *Coeligena conradii* is not recognized by IOC, eBird or H&M, all of which treat it as a subspecies of Collared Inca *C. torquata*.

[17] **Vilcabamba Inca** *Coeligena eisenmanni* is not recognized by IOC, eBird or H&M, all of which treat it as a subspecies of Collared Inca *C. torquata*.

[18] **Gould's Inca** *Coeligena inca* is not recognized by IOC, eBird or H&M, all of which treat it as a subspecies of Collared Inca *C. torquata*.

[19] **Huanuco Starfrontlet** *Coeligena dichroura* is not recognized by IOC, eBird or H&M, all of which treat it as a subspecies of what BirdLife call Bolivian Starfrontlet *C. violifer*, but use the name Violet-throated Starfrontlet.

[20] **Apurimac Starfrontlet** *Coeligena albicaudata* is not recognized by IOC, eBird or H&M, all of which treat it as a subspecies of what BirdLife call Bolivian Starfrontlet *C. violifer* but use the name Violet-throated Starfrontlet.

[21] **Cuzco Starfrontlet** *Coeligena osculans* is not recognized by IOC, eBird or H&M, all of which treat it as a subspecies of what BirdLife call Bolivian Starfrontlet *C. violifer*, but use the name Violet-throated Starfrontlet.

NT ↓ **White-tailed Starfrontlet** *Coeligena phalerata*

Monotypic 1,400–3,700 m
N **Colombia**

pp. 79, 99

EN ↓ **Glittering Starfrontlet** *C. orina*

Monotypic 2,500–3,500 m
W **Colombia**

pp. 56, 200

LC ↓ **Buff-winged Starfrontlet** *C. lutetiae*

2 sspp. 2,600–3,700 m
C Colombia to NW Peru

pp. 27, 64

EN ↓ **Perija Starfrontlet** [22] *C. consita*

Monotypic 2,550–3,100 m
NE Colombia/NW Venezuela

NOT ILLUSTRATED

LC ↓ **Golden-bellied Starfrontlet** *C. bonapartei*

Monotypic 2,150–3,000 m
NC **Colombia**

pp. 9, 13, 113

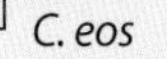

LC ↓ **Golden Starfrontlet** [23] *C. eos*

Monotypic 1,400–3,200 m
W **Venezuela**

NOT ILLUSTRATED

LC ? **Blue-throated Starfrontlet** *C. helianthea*

2 sspp. 1,900–3,300 m
NE Colombia/NW Venezuela; W Venezuela to C Colombia

pp. 54, 113, 213

Lafresnaya

1 species
p. 247 BL

LC = **Mountain Velvetbreast** *L. lafresnayi*

7 sspp. 1,800–3,700 m
W Venezuela to S Peru

p. 247

Ensifera

1 species
p. 247 BC

LC = **Sword-billed Hummingbird** *E. ensifera*

Monotypic 1,700–3,500 m
W Venezuela to C Bolivia

pp. 35, 144, 247

Pterophanes

1 species
p. 247 BR

LC = **Great Sapphirewing** *P. cyanopterus*

3 sspp. 2,600–3,700 m
NC Colombia to C Bolivia

pp. 26, 143, 247

Boissonneaua

3 species
p. 248 TL

LC ? **Buff-tailed Coronet** *B. flavescens*

2 sspp. 2,000–3,500 m
W Venezuela to N Ecuador

pp. 94, 151, 248

LC ? **Chestnut-breasted Coronet** *B. matthewsii*

Monotypic 1,200–3,000 m
SW Colombia to S Peru

pp. 44, 99, 152, 189

LC ↓ **Velvet-purple Coronet** *B. jardini*

Monotypic 700–2,300 m
WC Colombia to N Ecuador

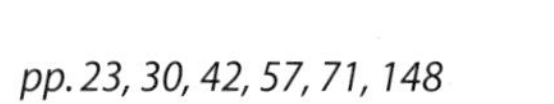

pp. 23, 30, 42, 57, 71, 148

Ocreatus

1 species
p. 248 TR

LC = **Booted Racket-tail** [24] *O. underwoodii*

8 sspp. 850–3,000 m
N Venezuela & N Colombia to C Bolivia

pp. 19, 137, 152, 214, 248

Urochroa [25]

2 species
p. 248 ML

LC ? **Rufous-gaped Hillstar** *U. bougueri*

Monotypic 800–2,800 m
WC Colombia to N Ecuador

p. 146

LC ? **White-tailed Hillstar** *U. leucura*

Monotypic 800–2,800 m
SW Colombia to NW Peru

p. 248

Urosticte

2 species
p. 248 MR

LC ↓ **Purple-bibbed Whitetip** *U. benjamini*

Monotypic 700–1,600 m
WC Colombia to N Ecuador

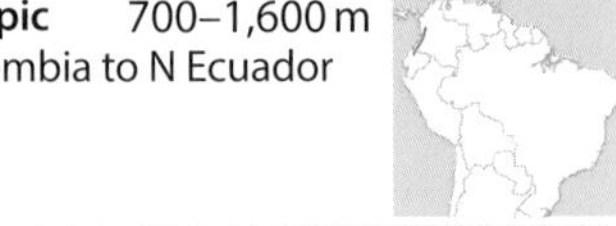

p. 150

LC ↓ **Rufous-vented Whitetip** *U. ruficrissa*

Monotypic 1,200–2,400 m
SW Colombia to NW Peru

p. 248

Heliodoxa

10 species
p. 248 BL

LC ↓ **Velvet-browed Brilliant** *H. xanthogonys*

2 sspp. 700–2,000 m
S Venezuela/N Brazil to W Guyana

NOT ILLUSTRATED

VU ↓ **Pink-throated Brilliant** *H. gularis*

Monotypic 350–1,100 m
SC Colombia to N Ecuador

NOT ILLUSTRATED

LC ↓ **Rufous-webbed Brilliant** *H. branickii*

Monotypic 700–1,550 m
C **Peru**

NOT ILLUSTRATED

LC ↓ **Black-throated Brilliant** *H. schreibersii*

Monotypic 400–1,300 m
NW Amazonia

p. 241

[22] **Perija Starfrontlet** *Coeligena consita* is not recognized by IOC, eBird or H&M, all of which treat it as a subspecies of Golden-bellied Starfrontlet *C. bonapartei*.

[23] **Golden Starfrontlet** *Coeligena eos* is not recognized by IOC, eBird or H&M, all of which treat it as a subspecies of Golden-bellied Starfrontlet *C. bonapartei*.

[24] **Booted Racket-tail** *Ocreatus underwoodii* is split into three species by IOC: Peruvian Racket-tail *O. peruanus* (E Ecuador to NE Peru); Rufous-booted Racket-tail *O. addae* (C & S Peru to C Bolivia), with 2 sspp.; and White-booted Racket-tail *O. underwoodii*, with 5 sspp.

[25] The two species of ***Urochroa*** are treated as one by H&M, named White-tailed Hillstar *U. bougueri*, with 2 sspp.

[26] **Black-breasted Brilliant** *Heliodoxa whitelyana* is not recognized by IOC, eBird or H&M, all of which treat it as a subspecies of Black-throated Brilliant *C. schreibersii*.

LC ↓ **Black-breasted Brilliant** [26] *H. whitelyana*

Monotypic 600–1,300 m
C & SE **Peru**

NOT ILLUSTRATED

LC ↓ **Gould's Brilliant** *H. aurescens*

Monotypic 100–1,000 m
Amazonia

p. 164

LC ? **Fawn-breasted Brilliant** *H. rubinoides*

3 sspp. 1,100–2,600 m
W Colombia to CS Peru

p. 151

LC ↓ **Green-crowned Brilliant** *H. jacula*

3 sspp. 300–2,000 m
Costa Rica to C Colombia; SW Colombia to S Ecuador

pp. 42, 47, 82, 98, 222

LC ? **Empress Brilliant** *H. imperatrix*

Monotypic 400–2,000 m
W Colombia to NW Ecuador

p. 150

LC ↓ **Violet-fronted Brilliant** *H. leadbeateri*

4 sspp. 400–2,400 m
N Venezuela to C Bolivia

p. 248

Clytolaema

1 species
p. 248 BR

LC ? **Brazilian Ruby** *C. rubricauda*

Monotypic 400–2,100 m
SE **Brazil**

pp. 23, 30, 50, 170, 248

Subfamily **Patagoninae**

Patagona

1 species
p. 249 TL

LC = **Giant Hummingbird** *P. gigas*

2 sspp. 2,000–4,000 m
SW Colombia to C Chile

pp. 13, 177, 232, 249

Subfamily **Trochilinae**

Clade Emeralds

Chlorostilbon

15 species
p. 249 TC

LC ↓ **Golden-crowned Emerald** *C. auriceps*

Monotypic 0–2,800 m
W /S **Mexico**

NOT ILLUSTRATED

LC = **Cozumel Emerald** *C. forficatus*

Monotypic 0–100 m
Mexico (Cozumel Is.)

NOT ILLUSTRATED

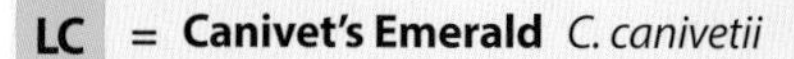

LC = **Canivet's Emerald** *C. canivetii*

3 sspp. 0–1,900 m
EC Mexico to C Costa Rica

NOT ILLUSTRATED

LC = **Garden Emerald** *C. assimilis*

Monotypic 0–1,500 m
S Costa Rica to E Panama

p. 13

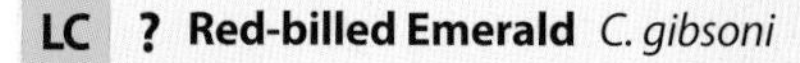

LC ? **Red-billed Emerald** *C. gibsoni*

3 sspp. 0–2,300 m
N Colombia/NW Venezuela to SW Colombia

p. 105

LC = **Blue-tailed Emerald** [27] *C. mellisugus*

8 sspp. 0–2,600 m
Bolivia/W Brazil to Venezuela to French Guiana/N Brazil

p. 265

LC ? **Chiribiquete Emerald** *C. olivaresi*

Monotypic 250–600 m
CS **Colombia**

NOT ILLUSTRATED

LC ? **Cuban Emerald** *C. ricordii*

Monotypic 0–1,300 m
Cuba & Bahamas

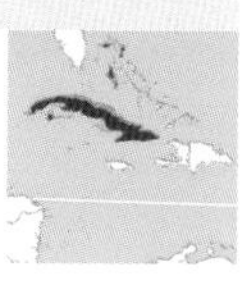

p. 123

Blue-tailed Emerald *Chlorostilbon mellisugus* ssp. *melanorhynchus* | W Ecuador

LC ? **Hispaniolan Emerald** *C. swainsonii*

Monotypic 300–2,500 m
Haiti/Dominican Republic (Hispaniola)

NOT ILLUSTRATED

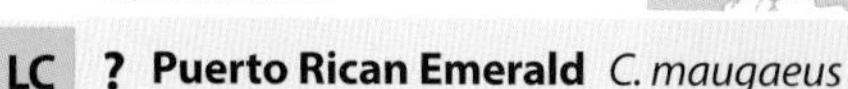

LC ? **Puerto Rican Emerald** *C. maugaeus*

Monotypic 0–800 m
Puerto Rico

NOT ILLUSTRATED

LC ? **Glittering-bellied Emerald** *C. lucidus*

4 sspp. 0–3,500 m
NE Brazil to E Bolivia/ N Argentina/Uruguay

p. 249

LC ? **Coppery Emerald** *C. russatus*

Monotypic 200–2,200 m
N Colombia/NW Venezuela

NOT ILLUSTRATED

LC ? **Narrow-tailed Emerald** *C. stenurus*

2 sspp. 750–3,000 m
NC Colombia to W Venezuela

NOT ILLUSTRATED

LC ? **Short-tailed Emerald** [28] *C. poortmani*

3 sspp. 800–2,200 m
CS Colombia to N Venezuela

NOT ILLUSTRATED

LC ? **Blue-chinned Emerald** *C. notatus*

3 sspp. 0–1,000 m
NC Colombia to SE Brazil; N & W Amazonia

p. 161

[27] **Blue-tailed Emerald** *Chlorostilbon mellisugus* is split into two species by IOC, eBird and H&M: subspecies *melanorhynchus* and *pumilus* being combined and called Western Emerald *C. melanorhynchus* (C Colombia to W & C Ecuador).

[28] **Short-tailed Emerald** *Chlorostilbon poortmani* is split into two species by IOC, eBird and H&M: subspecies *alice* being called Green-tailed Emerald *C. alice* (NW & N Venezuela).

Scaly-breasted Sabrewing *Campylopterus cuvierii* | Costa Rica

Rufous Sabrewing *Campylopterus rufus* | Guatemala

Cynanthus

4 species
p. 249 TR

LC = **Dusky Hummingbird** *C. sordidus*

Monotypic 900–2,200 m
S **Mexico**

NOT ILLUSTRATED

LC = **Broad-billed Hummingbird** *C. latirostris*

3 sspp. 0–2,200 m
S USA to C Mexico

pp. 24, 119, 249

LC = **Turquoise-crowned** [29] **Hummingbird** *C. doubledayi*

Monotypic 0–900 m
S **Mexico**

NOT ILLUSTRATED

NT = **Tres Marias Hummingbird** [30] *C. lawrencei*

Monotypic 0–600 m
Mexico (Tres Marías Is.)

NOT ILLUSTRATED

Cyanophaia

1 species
p. 249 MC

LC ↓ **Blue-headed Hummingbird** *C. bicolor*

Monotypic 0–1,000 m
Dominica & Martinique

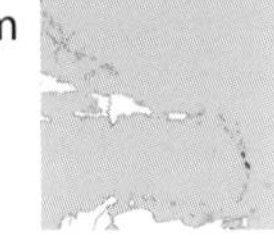

p. 249

Klais

1 species
p. 249 MR

LC ↓ **Violet-headed Hummingbird** *K. guimeti*

3 sspp. 0–1,900 m
E Honduras to NW Colombia; N Venezuela/N Colombia to C Bolivia

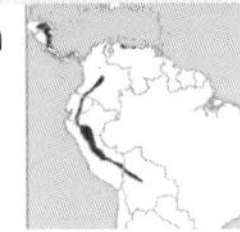

pp. 155, 249

Abeillia

1 species
p. 249 BC

LC ↓ **Emerald-chinned Hummingbird** *A. abeillei*

2 sspp. 1,000–2,200 m
CS Mexico to N Nicaragua

p. 249

Orthorhyncus

1 species
p. 249 BR

LC ? **Antillean Crested Hummingbird** *O. cristatus*

4 sspp. 0–500 m
Puerto Rico to Grenada & Barbados

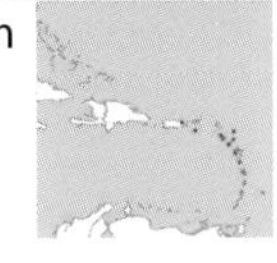

pp. 95, 127, 249

Stephanoxis [31]

2 species
p. 250 TL

LC ? **Green-crowned Plovercrest** *S. lalandi*

Monotypic 800–2,500 m
SE **Brazil**

pp. 168, 233, 250

LC ? **Violet-crowned Plovercrest** *S. loddigesii*

Monotypic 0–1,100 m
S Brazil to E Paraguay/ NE Argentina

NOT ILLUSTRATED

Aphantochroa

1 species
p. 250 TC

LC ? **Sombre Hummingbird** *A. cirrochloris*

Monotypic 0–700 m
E **Brazil**

pp. 171, 250

Anthocephala [32]

2 species
p. 250 TR

VU ↓ **Santa Marta Blossomcrown** *A. floriceps*

3 sspp. 600–1,700 m
N **Colombia**

p. 138

VU ↓ **Tolima Blossomcrown** *A. berlepschi*

3 sspp. 1,200–2,300 m
WC **Colombia**

p. 250

Campylopterus

13 species
p. 250 ML

LC = **Scaly-breasted Sabrewing** *C. cuvierii*

6 sspp. 0–1,200 m
SE Mexico/S Belize to E Panama; N Colombia

p. 266

LC ↓ **Wedge-tailed Sabrewing** [33] *C. curvipennis*

3 sspp. 60–1,500 m
EC Mexico to NE Honduras

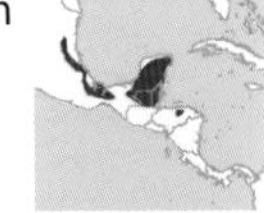

NOT ILLUSTRATED

LC ↓ **Grey-breasted Sabrewing** *C. largipennis*

2 sspp. 100–400 m
E Venezuela to N Brazil to Amazonia to SW Colombia

p. 165

VU ↓ **Dry-forest Sabrewing** [34] *C. calcirupicola*

Monotypic 450–900 m
EC **Brazil**

p. 239

NT ↓ **Diamantina Sabrewing** [35] *C. diamantinensis*

Monotypic 1,100–2,400 m
EC **Brazil**

NOT ILLUSTRATED

LC ? **Rufous Sabrewing** *C. rufus*

Monotypic 900–2,000 m
SE Mexico to El Salvador

p. 266

LC = **Rufous-breasted Sabrewing** *C. hyperythrus*

Monotypic 1,200–2,600 m
E Venezuela/W Guyana/ N Brazil

NOT ILLUSTRATED

LC ↓ **Violet Sabrewing** *C. hemileucurus*

2 sspp. 100–2,400 m
S Mexico to N Nicaragua; Costa Rica to S Panama

pp. 129, 221

NT ↓ **White-tailed Sabrewing** *C. ensipennis*

Monotypic 300–1,850 m
NE Venezuela & Tobago

p. 163

LC ? **Lazuline Sabrewing** *C. falcatus*

Monotypic 900–3,000 m
N Colombia/N Venezuela to N Ecuador

p. 250

CR ↓ **Santa Marta Sabrewing** *C. phainopeplus*

Monotypic 1,200–4,800 m
N **Colombia**

NOT ILLUSTRATED

NT ↓ **Napo Sabrewing** *C. villaviscensio*

Monotypic 900–1,700 m
SW Colombia to NW Peru

p. 22

LC ? **Buff-breasted Sabrewing** *C. duidae*

2 sspp. 1,200–2,400 m
S Venezuela to N Brazil

NOT ILLUSTRATED

Black-bellied Hummingbird *Eupherusa nigriventris* | Costa Rica

Eupetomena

1 species
p. 250 MC

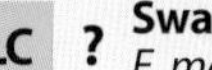

LC ? **Swallow-tailed Hummingbird** *E. macroura*

5 sspp. 0–1,700 m
S Suriname/N Brazil; NE Brazil to NE Argentina to SE Peru

pp. 160, 250

Eupherusa

4 species
p. 250 MR

LC ? **Stripe-tailed Hummingbird** *E. eximia*

2 sspp. 800–2,000 m
S Mexico to W Panama

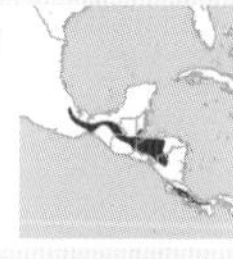

pp. 130, 250

EN ↓ **Oaxaca Hummingbird** *E. cyanophrys*

Monotypic 700–2,600 m
S **Mexico**

p. 196

NT ↓ **White-tailed Hummingbird** *E. poliocerca*

Monotypic 900–2,400 m
S **Mexico**

NOT ILLUSTRATED

LC ? **Black-bellied Hummingbird** *E. nigriventris*

Monotypic 900–2,000 m
W Costa Rica to W Panama

pp. 25, 59, 267

Elvira

2 species
p. 250 BL

LC ↓ **White-tailed Emerald** *E. chionura*

Monotypic 700–2,000 m
S Costa Rica to W Panama

NOT ILLUSTRATED

LC = **Coppery-headed Emerald** *E. cupreiceps*

Monotypic 700–1,500 m
W **Costa Rica**

pp. 132, 250

Microchera

1 species
p. 250 BC

LC ↓ **Snowcap** *M. albocoronata*

2 sspp. 300–1,600 m
E Honduras to W Panama

pp. 56, 84, 132, 230, 250

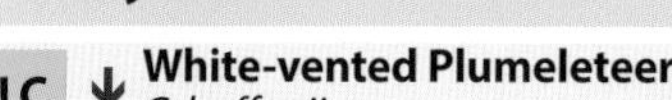

Bronze-tailed Plumeleteer *Chalybura urochrysia* | Costa Rica

Chalybura

2 species
p. 250 BR

LC ↓ **White-vented Plumeleteer** *C. buffonii*

4 sspp. 0–2,000 m
C Panama to N Venezuela/ SW Colombia

p. 250

LC ↓ **Bronze-tailed Plumeleteer** *C. urochrysia*

4 sspp. 0–900 m
E Honduras to NW Ecuador; SW Ecuador/NW Peru

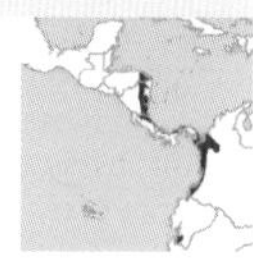

p. 267

[29] **Turquoise-crowned Hummingbird** *Cynanthus doubledayi* is not recognized by eBird or H&M, both of which treat it as a subspecies of Broad-billed Hummingbird *C. latirostris*; IOC do recognize but use the name Doubleday's Hummingbird.

[30] **Tres Marias Hummingbird** *Cynanthus lawrencei* is not recognized by IOC, eBird or H&M, all of which treat it as a subspecies of Broad-billed Hummingbird *C. latirostris*.

[31] The two species of ***Stephanoxis*** are treated as one by H&M, named Plovercrest *S. lalandi*, with 2 sspp.

[32] The two species of ***Anthocephala*** are treated as one by H&M, named Blossomcrown *A. floriceps*, with 2 sspp.

[33] **Wedge-tailed Sabrewing** *Campylopterus curvipennis* is split into three species and included in a different genus by IOC: Curve-winged Sabrewing *Pampa curvipennis* (EC Mexico), Wedge-tailed Sabrewing *P. pampa* (E Mexico to NE Honduras), and Long-tailed Sabrewing *P. excellens* (S Mexico). eBird and H&M spilt into two species: Wedge-tailed Sabrewing *P./C. curvipennis* and Long-tailed Sabrewing *P./C. excellens* respectively (S Mexico).

[34] **Dry-forest Sabrewing** *Campylopterus calcirupicola* is not (yet) recognized by H&M (newly described species); IOC and eBird use the name Outcrop Sabrewing.

[35] **Diamantina Sabrewing** *Campylopterus diamantinensis* is not recognized by H&M, which treats it as a subspecies of Grey-breasted Sabrewing *C. largipennis*.

Thalurania
5 species *p. 225* TL

VU ↓ **Mexican Woodnymph** *T. ridgwayi*

Monotypic 200–1,200 m
W **Mexico**

NOT ILLUSTRATED

LC ↓ **Crowned Woodnymph** *T. colombica*

8 sspp. 0–2,100 m
E Guatemala/S Belize to W Venezuela/NE Peru

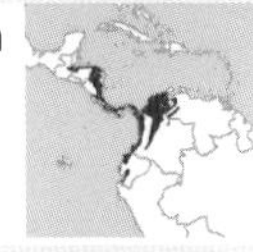

pp. 26, 31, 71, 77, 212

LC ? **Fork-tailed Woodnymph** *T. furcata*

13 sspp. 0–2,300 m
NE Venezuela/NE Brazil/E Ecuador/N Argentina

p. 160

EN ↓ **Long-tailed Woodnymph** *T. watertonii*

Monotypic 0–700 m
NE **Brazil**

p. 206

LC ? **Violet-capped Woodnymph** *T. glaucopis*

Monotypic 0–900 m
SE Brazil/N Uruguay/NE Argentina/E Paraguay

pp. 171, 251

White-bellied Hummingbird *Amazilia chionogaster* | Bolivia

Rufous-tailed Hummingbird *Amazilia tzacatl* | Colombia

Taphrospilus
1 species *p. 251* TC

LC ↓ **Many-spotted Hummingbird** *T. hypostictus*

Monotypic 400–1,860 m
SW Colombia to NW Peru; SC Peru to C Bolivia

p. 251

Leucochloris
1 species *p. 251* TR

LC ? **White-throated Hummingbird** *L. albicollis*

Monotypic 0–1,000 m
SE Brazil/E Argentina/E ParaguayE Bolivia

p. 251

Leucippus
4 species *p. 251* ML

LC ↓ **Buffy Hummingbird** *L. fallax*

Monotypic 0–550 m
NE Colombia to NE Venezuela

p. 162

LC ? **Tumbes Hummingbird** *L. baeri*

Monotypic 0–1,300 m
SE Ecuador to NW Peru

pp. 156, 251

LC ? **Spot-throated Hummingbird** *L. taczanowskii*

Monotypic 350–2,800 m
NW to WC **Peru**

NOT ILLUSTRATED

LC ? **Olive-spotted Hummingbird** *L. chlorocercus*

Monotypic 0–400 m
Amazonia rivers of E Ecuador & N Peru to NW Brazil

NOT ILLUSTRATED

Amazilia
44 species *p. 251* BL

LC ? **White-bellied Hummingbird** *A. chionogaster*

2 sspp. 400–2,800 m
NW Peru to N Argentina

p. 268

LC ? **Green-and-white Hummingbird** *A. viridicauda*

Monotypic 450–2,500 m
C&CS **Peru**

NOT ILLUSTRATED

LC ↓ **Rufous-tailed Hummingbird** *A. tzacatl*

5 sspp. 0–2,500 m
S Mexico to W Venezuela/S Ecuador

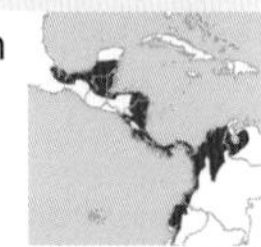

pp. 101, 111, 268

NT ↓ **Chestnut-bellied Hummingbird** *A. castaneiventris*

Monotypic 200–2,200 m
NC **Colombia**

p. 200

LC ↑ **Buff-bellied Hummingbird** *A. yucatanensis*

3 sspp. 0–1,250 m
S USA to Belize

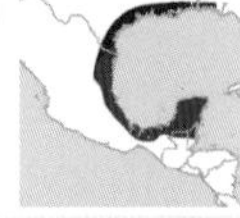

p. 119

LC = **Cinnamon Hummingbird** *A. rutila*

4 sspp. 0–1,700 m
NW Mexico & Yucatán Peninsula to W Costa Rica

p. 130

LC ? **Amazilia Hummingbird** *A. amazilia*

6 sspp. 0–2,500 m
NW Ecuador to SW Peru

p. 156

LC ? **Plain-bellied Emerald** *A. leucogaster*

2 sspp. 0–300 m
NE Venezuela to N Brazil; C Guyana; E Brazil

p. 163

LC ? **Versicolored Emerald** *A. versicolor*

6 sspp. 0–1,250 m
E Colombia to W Suriname; N&E Brazil to NW Argentina/E Paraguay/C Bolivia

p. 171

LC ? **White-chested Emerald** *A. brevirostris*

3 sspp. 0–500 m
N & C Venezuela/Trinidad/N Brazil/N French Guiana

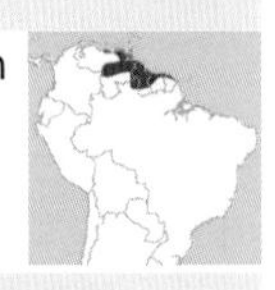

NOT ILLUSTRATED

LC ? **Andean Emerald** *A. franciae*

3 sspp. 300–2,100 m
W Colombia to WC Peru

pp. 25, 148

LC ↓ **White-bellied Emerald** *A. candida*

3 sspp. 0–1,500 m
EC Mexico to Nicaragua

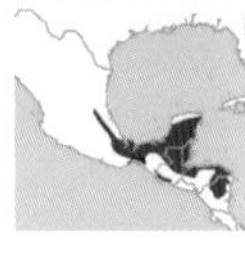

NOT ILLUSTRATED

VU ↓ **Honduran Emerald** *A. luciae*

Monotypic 200–500 m
N **Honduras**
p. 196

EN ↓ **Mangrove Hummingbird** *A. boucardi*

Monotypic 0–100 m
Costa Rica (Pacific coast)
p. 197

LC ↓ **Blue-chested Hummingbird** *A. amabilis*

Monotypic 0–1,000 m
Nicaragua to C Colombia/ S Ecuador
p. 8

LC ↓ **Charming Hummingbird** *A. decora*

Monotypic 300–1,200 m
S Costa Rica to SW Panama
NOT ILLUSTRATED

LC ↓ **Azure-crowned Hummingbird** *A. cyanocephala*

2 sspp. 0–2,500 m
EC Mexico to Nicaragua
NOT ILLUSTRATED

LC = **Berylline Hummingbird** *A. beryllina*

5 sspp. 600–2,200 m
NW Mexico to Honduras
NOT ILLUSTRATED

LC ↓ **Blue-tailed Hummingbird** *A. cyanura*

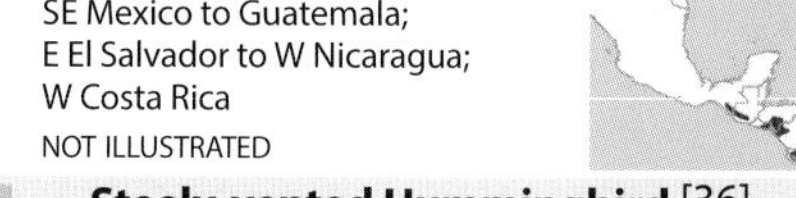

3 sspp. 0–1,500 m
SE Mexico to Guatemala;
E El Salvador to W Nicaragua;
W Costa Rica
NOT ILLUSTRATED

LC ↓ **Steely-vented Hummingbird** [36] *A. saucerottei*

4 sspp. 0–2,000 m
SW Nicaragua to S Costa Rica;
N Colombia to W Venezuela;
W Colombia
p. 32

LC = **Indigo-capped Hummingbird** *A. cyanifrons*

Monotypic 400–2,000 m
C **Colombia**
p. 269

CR = **Guanacaste Hummingbird** [37] *A. alfaroana*

Monotypic 600–650 m
W **Costa Rica**
NOT ILLUSTRATED

LC = **Violet-crowned Hummingbird** *A. violiceps*

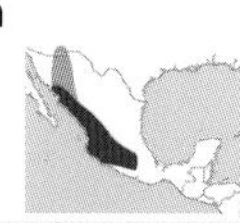

2 sspp. 0–2,400 m
S USA to S Mexico
p. 120

LC ↓ **Green-fronted Hummingbird** *A. viridifrons*

2 sspp. 60–1,400 m
S Mexico; SE Mexico to W Guatemala
NOT ILLUSTRATED

LC ↓ **Cinnamon-sided Hummingbird** [38] *A. wagneri*

Monotypic 200–900 m
S **Mexico**
NOT ILLUSTRATED

LC ? **Glittering-throated Emerald** *A. fimbriata*

7 sspp. 0–1,100 m
N Venezuela/NE Brazil/
E Ecuador/SE Brazil
pp. 30, 154

LC ↓ **Sapphire-spangled Emerald** *A. lactea*

2 sspp. 0–1,400 m
SE Venezuela/N Brazil;
SE Brazil
p. 51

LC ↓ **Spot-vented Emerald** [39] *A. bartletti*

Monotypic 0–1,400 m
E Peru/SW Brazil/N Bolivia
NOT ILLUSTRATED

LC ? **Purple-chested Hummingbird** *A. rosenbergi*

Monotypic 0–500 m
W Colombia to NW Ecuador
p. 269

LC = **Snowy-bellied Hummingbird** *A. edward*

4 sspp. 0–1,800 m
SE Costa Rica to E Panama
p. 88

LC ↓ **Green-bellied Hummingbird** *A. viridigaster*

2 sspp. 200–2,100 m
W Venezuela to C Colombia
NOT ILLUSTRATED

LC ↓ **Copper-tailed Hummingbird** [40] *A. cupreicauda*

4 sspp. 500–2,000 m
SE Venezuela/W Guyana/
N Brazil
NOT ILLUSTRATED

Indigo-capped Hummingbird *Amazilia cyanifrons* | Colombia

Purple-chested Hummingbird *Amazilia rosenbergi* | Colombia

[36] **Steely-vented Hummingbird** *Amazilia saucerottei* is split into two species and included in a different genus by IOC and eBird, ssp. *hoffmanni* being called Blue-vented Hummingbird *Saucerottia hoffmanni* (W Nicaragua to C Costa Rica).

[37] **Guanacaste Hummingbird** *Amazilia alfaroana* is not recognized by IOC or eBird, and is treated by H&M as a subspecies of Indigo-capped Hummingbird *A. cyanifrons*.

[38] **Cinnamon-sided Hummingbird** *Amazilia wagneri* is not recognized by eBird or H&M, which treat it as a subspecies of Green-fronted Hummingbird *A. viridifrons*, but by eBird in a different genus, *Leucolia*.

[39] **Spot-vented Emerald** *Amazilia bartletti* is not recognized by IOC, eBird or H&M, which treat it as a subspecies of Sapphire-spangled Emerald *A. viridifrons,* but by IOC and eBird in a different genus, *Chionomesa*.

[40] **Copper-tailed Hummingbird** *Amazilia cupreicauda* is not recognized by IOC, eBird or H&M, which treat it as a subspecies of Green-bellied Hummingbird *A. viridigaster*, but by IOC and eBird in a different genus, *Saucerottia*.

Humboldt's Hummingbird *Amazilia humboldtii* | Colombia

Shining-green Hummingbird *Amazilia goudoti* | Colombia

LC ? Copper-rumped Hummingbird *Amazilia tobaci*

7 sspp. 0–2,000 m
N Venezuela; Trinidad and Tobago

p. 162

LC ? Rufous-throated Hummingbird *A. sapphirina*

Monotypic 0–900 m
E Ecuador/N Peru to French Guiana/N Brazil to NE Bolivia; SE Brazil; E Paraguay/S Brazil/NE Argentina

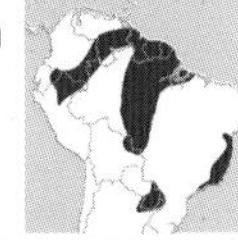

p. 251

LC ↓ Humboldt's Hummingbird *A. humboldtii*

Monotypic 0–200 m
SE Panama to NE Ecuador

p. 270

LC ↓ Gray's Hummingbird *A. grayi*

Monotypic 600–2,200 m
WC Colombia to N Ecuador

NOT ILLUSTRATED

LC = Golden-tailed Sapphire *A. oenone*

3 sspp. 0–1,600 m
Trinidad/N Venezuela to C Bolivia

pp. 9, 155

LC ↓ Violet-bellied Hummingbird *A. julie*

3 sspp. 0–1,800 m
C Panama to C Colombia; SW Colombia to NW Peru

pp. 36, 47, 76

LC ↑ Blue-throated Goldentail *A. eliciae*

2 sspp. 0–1,100 m
SE Mexico to NW Colombia

p. 52

LC ? White-chinned Sapphire *A. cyanus*

5 sspp. 0–1,250 m
N Colombia/W & N Venezuela; E Venezuela/N Brazil/C Bolivia/N Peru; E & SE Brazil

p. 29

LC ↑ Gilded Hummingbird *A. chrysura*

Monotypic 200–1,000 m
N Bolivia/SE Brazil/E Argentina

NOT ILLUSTRATED

LC = Sapphire-throated Hummingbird *A. coeruleogularis*

3 sspp. 0–100 m
SW Panama to N Colombia

p. 70

EN ↓ Sapphire-bellied Hummingbird *A. lilliae*

Monotypic 0–100 m
N **Colombia**

pp. 190, 199

LC ↑ Shining-green Hummingbird *A. goudoti*

4 sspp. 0–1,600 m
N Colombia/W Venezuela to SW Colombia

p. 270

Trochilus [41]

2 species
p. 251 BR

LC ? Red-billed Streamertail *T. polytmus*

Monotypic 0–1,600 m
Jamaica

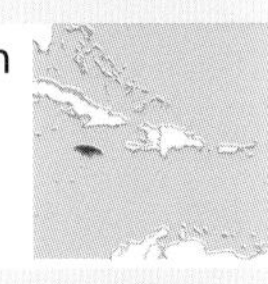

pp. 125, 251

LC ? Black-billed Streamertail *T. scitulus*

Monotypic 0–1,100 m
E **Jamaica**

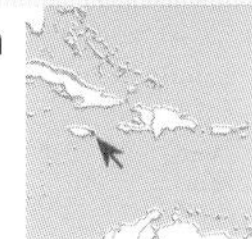

pp. 125, 226, 271

Goethalsia

1 species
p. 252 TL

NT ? Pirre Hummingbird *G. bella*

Monotypic 500–1,500 m
SE Panama/NW Colombia

pp. 197, 252

Goldmania

1 species
p. 252 TC

NT ↓ Violet-capped Hummingbird *G. violiceps*

Monotypic 600–1,200 m
C Panama to NW Colombia

p. 252

Basilinna

2 species
p. 252 TR

LC = Xantus's Hummingbird *B. xantusii*

Monotypic 150–1,500 m
S Baja California, **Mexico**

p. 121

LC ? White-eared Hummingbird *B. leucotis*

3 sspp. 1,200–3,500 m
NW Mexico to W Nicaragua

pp. 108, 121, 252

[41] The two species of ***Trochilus*** are treated as one by eBird, named Streamertail *T. polytmus*, with 2 sspp.

[42] **Magnificent Hummingbird** *Eugenes fulgens* is split into two species by IOC and eBird: ssp. *spectabilis* called Talamanca Hummingbird *E. spectabilis* (C Costa Rica to W Panama) and *E. fulgens* called Rivoli's Hummingbird (SW USA to W Nicaragua).

Clade Mountain-gems

Sternoclyta
1 species
p. 252 MR

LC ? **Violet-chested Hummingbird** *S. cyanopectus*
Monotypic 0–2,000 m
N Venezuela to NC Colombia

pp. 252

Hylonympha
1 species
p. 252 BL

EN ↓ **Scissor-tailed Hummingbird** *H. macrocerca*
Monotypic 500–1,200 m
NE **Venezuela**

pp. 198, 252

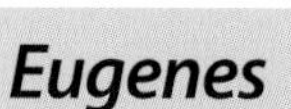

Eugenes
1 species
p. 252 BC

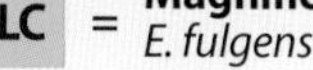

LC = **Magnificent Hummingbird** [42] *E. fulgens*
2 sspp. 1,500–3,000 m
SW USA to W Nicaragua;
C Costa Rica to W Panama

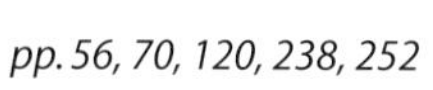

pp. 56, 70, 120, 238, 252

Panterpe
1 species
p. 252 BR

LC = **Fiery-throated Hummingbird** *P. insignis*
2 sspp. 1,600–3,200 m
W Costa Rica to W Panama

pp. 1, 67, 133, 252

Black-billed Streamertail *Trochilus scitulus* | Jamaica

Heliomaster
4 species
p. 253 TL

LC ↓ **Plain-capped Starthroat** *H. constantii*
3 sspp. 0–1,500 m
NW Mexico to E Costa Rica

NOT ILLUSTRATED

LC ↓ **Long-billed Starthroat** *H. longirostris*
3 sspp. 0–1,500 m
S Mexico to Trinidad and Tobago;
French Guiana/Amazonia/W Ecuador

pp. 253, 271

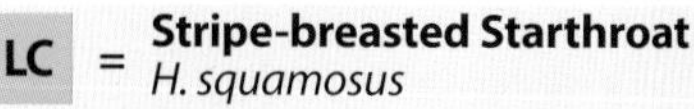

LC = **Stripe-breasted Starthroat** *H. squamosus*
Monotypic 0–800 m
Brazil

pp. 166, 227

LC ↓ **Blue-tufted Starthroat** *H. furcifer*
Monotypic 0–1,000 m
SC Brazil/N Uruguay/
N Argentina/W Bolivia

p. 73

Lampornis
8 species
p. 253 TC

LC ↓ **Green-throated Mountain-gem** *L. viridipallens*
4 sspp. 900–2,700 m
SE Mexico to C Honduras/
N El Salvador

p. 133

LC ↓ **Green-breasted Mountain-gem** *L. sybillae*
Monotypic 750–2,400 m
C Honduras to W Nicaragua

NOT ILLUSTRATED

LC ↓ **Amethyst-throated Hummingbird** *L. amethystinus*
5 sspp. 900–3,000 m
se Venezuela to n Brazil;
w Brazil

p. 121

LC = **Blue-throated Hummingbird** *L. clemenciae*
3 sspp. 1,500–3,500 m
SW USA to S Mexico

p. 120

LC ? **White-bellied Mountain-gem** *L. hemileucus*
Monotypic 700–1,400 m
W Costa Rica to NC Panama

p. 13

Long-billed Starthroat *Heliomaster longirostris* | Panama

LC ↓ **Purple-throated Mountain-gem** *L. calolaemus*
3 sspp. 800–2,500 m
SE Nicaragua to NC Panama

p. 133

LC ? **Grey-tailed Mountain-gem** *L. cinereicauda*
Monotypic 1,800–3,000 m
E **Costa Rica**

p. 83

LC ? **White-throated Mountain-gem** *L. castaneoventris*
Monotypic 1,800–3,000 m
W **Panama**

pp. 50, 253

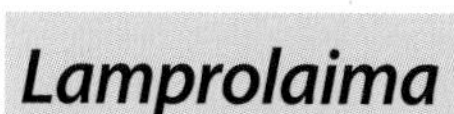

Lamprolaima
1 species
p. 253 TR

LC ? **Garnet-throated Hummingbird** *L. rhami*
Monotypic 1,200–2,800 m
EC Mexico to E Honduras

p. 253

Gorgeted Woodstar *Chaetocercus heliodor* | Colombia

Clade Bees

Myrtis
1 species *p. 253* ML

LC = **Purple-collared Woodstar** *M. fanny*

2 sspp. 1,200–2,800 m
N Ecuador to S Peru

p. 253

Eulidia
1 species *p. 253* BL

CR ↓ **Chilean Woodstar** *E. yarrellii*

Monotypic 200–1,840 m
N **Chile**

pp. 157, 204, 253

Rhodopis
1 species *p. 253* BC

LC = **Oasis Hummingbird** *R. vesper*

3 sspp. 0–3,000 m
NW Peru to C Chile

pp. 109, 157, 253

Thaumastura
1 species *p. 253* BR

LC = **Peruvian Sheartail** *T. cora*

Monotypic 0–2,400 m
SW Ecuador to N Chile

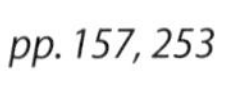

pp. 157, 253

Chaetocercus
6 species *p. 254* TL

LC = **White-bellied Woodstar** *C. mulsant*

Monotypic 1,500–3,300 m
NC Colombia to C Bolivia

pp. 13, 22, 150

NT ↓ **Little Woodstar** *C. bombus*

Monotypic 0–3,000 m
SW Colombia to NW Peru

NOT ILLUSTRATED

LC = **Gorgeted Woodstar** *C. heliodor*

2 sspp. 1,200–3,000 m
W Venezuela to C Ecuador

pp. 254, 272

LC = **Santa Marta Woodstar** *C. astreans*

Monotypic 800–2,200 m
N **Colombia**

NOT ILLUSTRATED

VU ↓ **Esmeraldas Woodstar** *C. berlepschi*

Monotypic 0–675 m
W **Ecuador**

p. 201

LC = **Rufous-shafted Woodstar** *C. jourdanii*

3 sspp. 900–3,000 m
NE Colombia/NW Venezuela; NC Colombia to N Venezuela; NE Venezuela & Trinidad

NOT ILLUSTRATED

Myrmia
1 species *p. 254* TC

LC = **Short-tailed Woodstar** *M. micrura*

Monotypic 0–200 m
WC Ecuador to NE Peru

pp. 157, 254

Microstilbon
1 species *p. 254* TR

LC = **Slender-tailed Woodstar** *M. burmeisteri*

Monotypic 460–2,900 m
C Bolivia to N Argentina

p. 254

Calliphlox
3 species *p. 254* MR

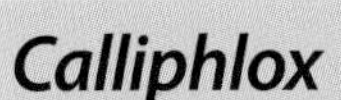

LC ↓ **Amethyst Woodstar** *C. amethystina*

Monotypic 0–1,500 m
W Venezuela/E Colombia to NE Brazil to SE Brazil/NE Argentina/E Paraguay to NE Ecuador/SW Colombia

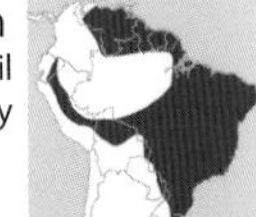

pp. 100, 104, 161

LC ↓ **Magenta-throated Woodstar** *C. bryantae*

Monotypic 700–1,900 m
W Costa Rica to C Panama

NOT ILLUSTRATED

LC = **Purple-throated Woodstar** *C. mitchellii*

Monotypic 700–2,200 m
E Panama; WC Colombia to N Ecuador; SW Ecuador

pp. 71, 254

Calliope Hummingbird *Selasphorus calliope* | Canada

Doricha
2 species
p. 254 BL

LC ↓ **Slender Sheartail** *D. enicura*

Monotypic 1,000–3,000 m
SE Mexico to W Honduras/
N El Salvador

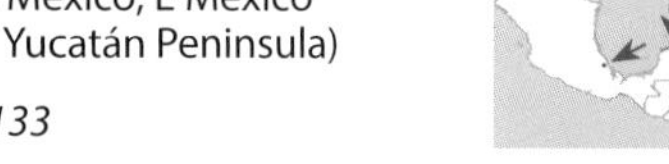

p. 254

NT ↓ **Mexican Sheartail** *D. eliza*

Monotypic 0–1,450 m
EC Mexico; E Mexico
(N Yucatán Peninsula)

p. 133

Tilmatura
1 species
p. 254 BC

LC = **Sparkling-tailed Woodstar**
T. dupontii

Monotypic 500–2,500 m
W Mexico to W Nicaragua

p. 254

Calothorax
2 species
p. 254 BR

LC = **Lucifer Hummingbird** *C. lucifer*

Monotypic 1,200–2,360 m
SW USA to S Mexico

pp. 120, 254

LC ↓ **Beautiful Hummingbird** *C. pulcher*

Monotypic 1,000–2,200 m
S **Mexico**

NOT ILLUSTRATED

Volcano Hummingbird *Selasphorus flammula* | Panama

Nesophlox
2 species
p. 255 TL

LC = **Bahama Hummingbird** *N. evelynae*

Monotypic 0–50 m
Bahamas; Turks and Caicos
Islands

p. 255

LC = **Lyre-tailed Hummingbird** *N. lyrura*

Monotypic 0–37 m
Bahamas

NOT ILLUSTRATED

Mellisuga
2 species
p. 255 TC

LC = **Vervain Hummingbird** *M. minima*

2 sspp. 0–1,600 m
Jamaica; Haiti/Dominican
Republic (Hispaniola); Puerto
Rico

p. 124

NT ↓ **Bee Hummingbird** *M. helenae*

Monotypic 0–1,200 m
Cuba

pp. 123, 195, 209, 255

Calypte
2 species
p. 255 TR

LC ↑ **Anna's Hummingbird** *C. anna*

Monotypic 0–1,800 m
NW USA (Alaska)
to NW Mexico

pp. 28, 31, 38, 69, 118, 210, 255

LC ↑ **Costa's Hummingbird** *C. costae*

Monotypic 0–1,000 m
SW USA to W Mexico

p. 119

Archilochus
2 species
p. 255 BL

LC ↑ **Black-chinned Hummingbird**
A. alexandri

Monotypic 0–2,000 m
SW Canada to S Mexico

pp. 110, 118

LC ↑ **Ruby-throated Hummingbird**
A. colubris

Monotypic 0–1,900 m
S Canada to C Panama

pp. 118, 175, 211, 229, 255

Selasphorus
7 species
p. 255 BC

LC = **Broad-tailed Hummingbird**
S. platycercus

Monotypic 1,500–2,500 m
W USA to S Guatemala

p. 118

LC ↑ **Calliope Hummingbird** *S. calliope*

Monotypic 185–3,400 m
SW Canada to S Mexico

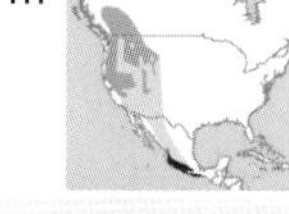

pp. 119, 272

NT ↓ **Rufous Hummingbird** *S. rufus*

Monotypic 0–3,000 m
NW USA (Alaska)/SW Canada
to S Mexico

pp. 119, 175, 195

LC ↑ **Allen's Hummingbird** *S. sasin*

2 sspp. 0–2,900 m
W **USA** to SC Mexico

pp. 17, 119

LC = **Volcano Hummingbird** *S. flammula*

3 sspp. 1,850–3,500 m
C Costa Rica to W Panama

p. 273

LC ↓ **Scintillant Hummingbird** *S. scintilla*

Monotypic 900–2,450 m
W Costa Rica to W Panama

pp. 255

EN ↓ **Glow-throated Hummingbird**
S. ardens

Monotypic 750–1,850 m
NC **Panama**

NOT ILLUSTRATED

Atthis
2 species
p. 255 BR

LC = **Bumblebee Hummingbird**
A. heloisa

2 sspp. 1,500–3,000 m
W/EC/S **Mexico**

pp. 88, 255

LC = **Wine-throated Hummingbird**
A. ellioti

2 sspp. 1,500–3,500 m
SE Mexico to SE Honduras

p. 61

Appendix 1 | Misnamed Hummingbirds

There are 30 hummingbirds that have been included as 'species' in published lists but are now known to be the same as another named species, or are suspected or have been shown to be aberrant or immature forms of other species.

Cristina's Barbthroat *Threnetes cristinae*, described in 1975 from Espírito Santo, Brazil, is now considered to be the same as Pale-tailed Barbthroat subspecies *loehkeni*.

Grzimek's Barbthroat *Threnetes grzimeki*, described in 1973, is now considered to be an immature plumage of Rufous-breasted Hermit.

Long-tailed Barbthroat *Threnetes longicauda*, described in 1915 from the type specimen from Ceará, Brazil, is now considered to be the same species as Broad-tipped Hermit.

Maranhão Hermit *Phaethornis maranhaoensis*, described in 1968, is now considered to represent the adult male plumage of Cinnamon-throated Hermit.

Black-billed Hermit *Phaethornis nigrirostris*, described in 1973, is now considered to represent aberrant black-billed individuals of Scale-throated Hermit.

Heini's Hermit *Phaethornis apheles*, described in 1984 from "northern Peru", was subsequently considered to be probably the same as another species, Phaethornis zonura, which is now considered a subspecies of Grey-chinned Hermit.

Schluter's Hermit *Phaethornis fumosus*, described in 1901 from "Colombia", is now considered to represent melanistic individuals of Sooty-capped Hermit.

The '**Maranhão Hermit**' was originally described as species, *Phaethornis maranhaoensis*, but has since been shown to represent the adult male plumage of **Cinnamon-throated Hermit** *Phaethornis nattereri*. | Brazil

Sooty Hermit *Phaethornis fuliginosus*, described in 1901 from the type specimen from "Bogotá", is now considered to represent a melanistic individual of another *Phaethornis* (species uncertain).

Buckley's Violet-ear *Colibri buckleyi*, described in 1893 and known only from the type specimen from Misqui, Bolivia, is now considered to represent an aberrant Sparkling Violet-ear.

Hartert's Sunangel *Heliangelus dubius*, described in 1897 and known from two "Bogotá" specimens, is now considered possibly to represent a melanistic form of Amethyst-throated Sunangel.

Claudia's Sunangel *Heliangelus claudia*, described in 1895 and known from "Bogotá" specimens, is now considered to represent an aberrant Longuemare's Sunangel.

Sarayacu Sunangel *Heliangelus violicollis*, described in 1891 and known from two specimens from uncertain localities in Ecuador, is now considered to represent a color variant of Gorgeted Sunangel.

Simon's Sunangel *Heliangelus simoni*, described in 1892 from "Colombia" and subsequently treated as the same species as '*Heliangelus speciosus*', which is not currently recognized; this hummingbird is therefore of uncertain status.

Oberholser's Sunangel *Heliangelus prosantis*, described in 1905 from "Bogotá" and subsequently treated as the same 'species' as Rothschild's Sunangel (see *opposite page*), which is probably a hybrid sunangel *Heliangelus* sp. × puffleg *Eriocnemis* sp. or Purple-backed Thornbill; this hummingbird is therefore of uncertain status.

Great-crested Coquette *Lophornis regulus*, described in 1846 and known from Peru and Bolivia, is now considered to be the same species as Rufous-crested Coquette.

Dusky Coquette *Lophornis melaniae*, described in 1920 and known from one specimen from "Colombia", is now considered probably to be an aberrant individual or faded skin of Rufous-crested Coquette.

Balsas Metaltail *Metallura rubriginosa*, described in 1913 and known only from the type specimen from northern Peru, is now treated as the same species as Coppery Metaltail.

Berlepsch's Puffleg *Eriocnemis berlepschi*, described in 1897 and known only from the type specimen from "Bogotá", Colombia, is now considered to be the same as Glowing Puffleg subspecies *vestita*.

Blue-vented Puffleg *Eriocnemis chrysorama*, described in 1874 and known only from the type specimen from "Colombia", is now considered to represent a melanistic specimen of Coppery-bellied Puffleg.

Allied Inca *Coeligena assimilis*, described in 1879 from "Bogotá", Colombia, is generally considered to be the same species as Black Inca but may represent a valid subspecies.

Lesson's Woodstar *Calliphlox orthura*, described in 1831 from "Cayenne", French Guiana, possibly represents an immature male Amethyst Woodstar.

Brilliant Emerald *Chlorostilbon micans*, described in 1892 from the type specimen, which has no locality, is now considered to be probably an aberrant Short-tailed Emerald subspecies *alice*.

Berlepsch's Emerald *Chlorostilbon inexpectatus*, described in 1879 and known only from one specimen from "Bogotá", Colombia, is now considered probably to represent an aberrant Short-tailed Emerald.

Cabanis's Emerald *Chlorostilbon auratus*, described in 1860 and known only from the type specimen from "Peru", is now considered likely to be an aberrant Short-tailed Emerald (which is found only in the northern Andes, indicating that the locality is incorrect).

Elliot's Sapphire *Chlorestes subcaerulea*, described in 1874 and known only from the type specimen from "Bahía", Brazil, is now considered probably to represent an aberrant Blue-chinned Emerald.

Tschudi's Woodnymph *Thalurania tschudii*, described in 1860, is now considered to be the same as Fork-tailed Woodnymph subspecies *jelskii*.

Peruvian Hummingbird *Leucippus pallidus*, described in 1874 and known from central Peru, is now considered to be the same as White-bellied Hummingbird subspecies *chionogaster*.

Blue-spotted Hummingbird *Amazilia cyaneotincta*, described in 1909 and known from two "Bogotá", Colombia, specimens, is now considered probably to be an aberrant Glittering-throated Emerald or Blue-chested Hummingbird.

Elliot's Hummingbird *Amazilia lucida*, described in 1877 and known from one specimen from "Colombia", is now considered to be of uncertain status.

Dusky Emerald *Amazilia veneta*, described in 1921 and known only from the type specimen from "Bogotá", Colombia, is now considered perhaps to be a melanistic female Andean Emerald.

Appendix 2 | Named Hummingbird Hybrids

As explained in the section on *Hybridization* (*page 112*), hybrid hummingbirds have, in the past, often been described as a species, some of which are now known to account for 'lost' species. For the sake of completeness, the 34 hummingbirds that were so named are listed here.

Elliot's Topaz *Chrysolampis chlorolaemus*, described in 1870 from the type specimen labeled "New Grenada" (later thought to be "Bahía", Brazil), is now considered a hybrid Black-throated Mango × Ruby-topaz Hummingbird.

Bogotá Sunangel *Heliangelus zusii*, described in 2011 based on a single specimen collected in 1909, is now known to be a hybrid between a female Long-tailed Sylph and a male of another (uncertain) species (see *page 112*).

Olive-throated Sunangel *Heliangelus squamigularis*, described in 1871 from specimens labeled "Bogotá" and "Antioquia", Colombia, is now known to be a hybrid Amethyst-throated Sunangel × Coppery-bellied Puffleg (see also Green-throated Sunangel).

Green-throated Sunangel *Heliangelus speciosa*, described in 1891 from specimens labeled "Bogotá", Colombia, is now known to be a hybrid, most likely Amethyst-throated Sunangel × Coppery-bellied Puffleg (see also Olive-throated Sunangel).

Rothschild's Sunangel *Heliangelus rothschildi*, described in 1892 and known only from "Bogotá", Colombia, specimens, is probably a hybrid sunangel *Heliangelus* sp. × puffleg *Eriocnemis* sp. or Purple-backed Thornbill (see also Glistening Sunangel, *next column*).

Glistening Sunangel *Heliangelus luminosus*, described in 1878 and known only from "Colombia", is probably a hybrid sunangel *Heliangelus* sp. × puffleg *Eriocnemis* sp. or Purple-backed Thornbill (see also Rothschild's Sunangel).

Bearded Coquette *Lophornis insignibarbis*, described in 1890 from a single specimen from "Bogotá", Colombia, is possibly a hybrid Spangled Coquette × Butterfly Coquette.

Orton's Comet *Lesbia ortoni*, described in 1890 from "Quito Valley", Ecuador, is probably a hybrid Black-tailed Trainbearer × Purple-backed Thornbill.

Bourcier's Trainbearer *Lesbia eucharis*, described in 1848 and known from two males from Colombia, has had a varied taxonomic history but is now considered a hybrid Green-tailed Trainbearer × Black-tailed Trainbearer.

Purple-tailed Comet *Zodalia glyceria*, described in 1858 from Popayán, Colombia, has been shown to be a hybrid Black-tailed Trainbearer × Rainbow-bearded Thornbill.

Chillo Valley Comet *Zodalia thaumasta*, described in 1902 from just one specimen from Illalo, Ecuador, is currently treated as synonymous with Purple-tailed Thornbill (see *page 276*), and therefore probably a hybrid Long-tailed Sylph × Black-tailed Trainbearer.

Purple-tailed Thornbill *Chalcostigma purpureicauda*, described in 1898 from just one specimen from "Bogotá", Colombia, is now considered probably to be a hybrid Long-tailed Sylph × Black-tailed Trainbearer.

Nehrkorn's Sylph *Neolesbia nehrkorni*, described in 1887 and known from two "Bogotá", Colombia, specimens, is probably a hybrid Long-tailed Sylph × Purple-backed Thornbill, or Long-tailed Sylph × woodnymph *Thalurania* sp.

Amethyst-vented Puffleg *Eriocnemis ventralis*, described in 1891 and known only from the type specimen from "Bogotá", Colombia, is now considered probably to be a hybrid Glowing Puffleg × Coppery-bellied Puffleg.

Söderström's Puffleg *Eriocnemis soederstroemi*, described in 1926 and known only from one specimen from Volcán Pichincha, in western Ecuador, has been shown to be a hybrid Black-breasted Puffleg × Sapphire-vented Puffleg.

Isaacson's Puffleg *Eriocnemis isaacsonii*, described in 1845 and known from three "Bogotá", Colombia, specimens, is possibly a hybrid puffleg *Eriocnemis* sp. × sunangel *Heliangelus* sp.

Purple Inca *Coeligena purpurea*, described in 1845 and known from two specimens from the Popayán area of Colombia, has been shown to be a hybrid Black Inca × Bronzy Inca.

Lilac-fronted Starfrontlet *Coeligena traviesi*, described in 1867 and known from several "Bogotá", Colombia, specimens that show considerable individual variation, is probably a hybrid Collared Inca × Buff-winged Starfrontlet.

Lawrence's Starfrontlet *Coeligena lawrencei*, described in 1893 from a single "Bogotá", Colombia, specimen, is probably a hybrid Collared Inca × Mountain Velvetbreast.

Gould's Woodstar *Chaetocercus decorata*, described in 1860 from the type specimen from "Bogotá", Colombia, is a hybrid White-bellied Woodstar × Gorgeted Woodstar.

Hartert's Woodstar *Chaetocercus harterti*, described in 1901 and known only from the type specimen from the central Andes of Colombia, is probably a hybrid, but the parentage is uncertain.

Iridescent Emerald *Smaragdochrysis iridescens*, described in 1861 and known only from the type specimen from Nova Friburgo, Rio de Janeiro, Brazil, is regarded as a hybrid Glittering-bellied Emerald × Amethyst Woodstar.

Natterer's Emerald *Ptochoptera iolaima*, described in 1854 and known only from the type specimen from Ypanema, São Paulo, Brazil, is considered to be a probable hybrid 'Green-tailed Emerald' (which BirdLife International treats as a subspecies of Short-tailed Emerald) × Short-tailed Emerald, both of which are found only in the northern Andes (indicating that the locality is incorrect).

Blue-breasted Sapphire *Eucephala hypocyanea*, described in 1860 and known only from the type specimen from "Brazil", is probably either an aberrant Blue-chinned Emerald or a hybrid Blue-chinned Emerald × White-chinned Sapphire.

Black-bellied Woodnymph *Eucephala scapulata*, described in 1861 and known only from the type specimen from French Guiana, is suspected to be a hybrid Fork-tailed Woodnymph × Blue-chinned Emerald.

Reeve's Woodnymph *Eucephala caeruleolavata*, described in 1860 and known only from the type specimen from southeast Brazil, is generally assumed to be a hybrid of some sort, but the parentage is uncertain.

Green-headed Woodnymph *Eucephala chlorocephala*, described in 1854 and known only from the type specimen from southeast Brazil, is likely a hybrid of uncertain parentage but may be the same as Gould's Woodnymph (see *below*).

Lerch's Woodnymph *Thalurania lerchi*, described in 1872 and known only from "Bogotá", Colombia, is suspected to be a hybrid Fork-tailed Woodnymph × Golden-tailed Sapphire.

Gould's Woodnymph *Augasma smaragdinea*, described in 1860 and known from six specimens from Nova Friburgo, Rio de Janeiro, Brazil, is generally assumed to be a hybrid of some sort, but the parentage is uncertain (see Green-headed Woodnymph, *above*).

Simon's Woodnymph *Augasma chlorophana*, described in 1897 and known only from the type specimen from "Bahía", Brazil, is generally assumed to be a hybrid of some sort, but the parentage is uncertain.

Berlioz's Woodnymph *Augasma cyaneoberyllina*, described in 1965 and known from two specimens from "Bahía", Brazil, is generally assumed to be a hybrid of some sort, but the parentage is uncertain.

Reichenbach's Whitethroat *Leucochloris malvina*, described in 1855 and known only from the type specimen from "Brazil", is possibly a hybrid White-throated Hummingbird × Golden-tailed Sapphire.

Táchira Emerald *Amazilia distans*, described in 1966 from Venezuela, is now known to be a hybrid White-chinned Sapphire × Glittering-throated Emerald.

Flame-rumped Sapphire *Hylocharis pyropygia*, described in 1881 and known from five specimens from "Bahía", Brazil, is probably a hybrid White-chinned Sapphire × Glittering-bellied Emerald, but could possibly be a valid species.

Further Reading and Sources of Useful Information

Books

Selected books specifically about hummingbirds that are currently in print, and key reference works.

Dunn, J. 2021. ***The Glitter in the Green: In Search of Hummingbirds***. Basic Books.

Gates, L. & Gates, T. 2007. ***Enjoying Hummingbirds In the Wild and in Your Yard***. Stackpole Books.

del Hoyo, J. 2020. ***All the Birds of the World***. Lynx Edicions.

del Hoyo, J., Collar, N. J., Christie, D., Elliott, A. & Fishpool, L. D. C. 2014. ***HBW and BirdLife International Illustrated Checklist of the Birds of the World***. Vol. 1. Lynx Edicions.

del Hoyo, J., Elliott, A. & Sargantal, J. 1999. ***Handbook of the Birds of the World***. Vol. 5. Lynx Edicions.

Kricher, J. C. 2017. ***The New Neotropical Companion***. Princeton University Press.

Orenstein, R. 2018. ***Hummingbirds***. Firefly Books.

Shewey, J. 2021. ***The Hummingbird Handbook: Everything You Need to Know about These Fascinating Birds***. Timber Press.

Skutch, A. F. 1973. ***The Life of the Hummingbird***. Crown Publishers, Inc.

West, C. G. 2015. ***North American Hummingbirds: An Identification Guide***. University of New Mexico Press.

Williamson, S. L. 2002. ***Hummingbirds of North America***. Houghton Mifflin Harcourt.

Key papers

A list of the books and key papers used for reference in preparing this book is available as a downloadable PDF from the Princeton University Press website at https://press.princeton.edu/books/hardcover/9780691182124/hummingbirds ('Resources' tab). The following papers are of particular note, however, and images from the paper by Ksepka *et al*. have been reproduced on *page 11* with the permission of the authors.

Chen, A. & Field, D. J. 2020. Phylogenetic definitions for Caprimulgimorphae (Aves) and major constituent clades under the International Code of Phylogenetic Nomenclature. ***Vertebrate Zoology***, **70(4)**: 571–585. https://doi.org/10.26049/VZ70-4-2020-03

Ksepka, D. T., Clarke, J. A., Nesbitt, S. J., Kulp, F. B., & Grande, L. 2013. Fossil evidence of wing shape in a stem relative of swifts and hummingbirds (Aves, Pan-Apodiformes). ***Proceedings of the Royal Society B***: Biological Sciences, **280(1761)**, 20130580. https://royalsocietypublishing.org/doi/10.1098/rspb.2013.0580

Online resources and taxonomic lists

Avibase **– The World Bird Database:** https://avibase.bsc-eoc.org

BirdLife International Datazone: http://datazone.birdlife.org/home

Clements, J. F., Schulenberg, T. S., Iliff, M. J., Billerman, S. M., Fredericks, T. A., Gerbracht, J. A., Lepage, D., Sullivan, B. L. & Wood, C. L. 2021. ***The eBird/Clements Checklist of Birds of the World***: v2021. https://www.birds.cornell.edu/clementschecklist

Dickinson, E. C. & Remsen Jr., J. V. (Eds). 2013. ***The Howard and Moore Complete Checklist of the Birds of the World***, Fourth Edition, Vol. 1: Non-passerines. Aves Press. https://www.aviansystematics.org

eBird: https://ebird.org/home

Gill F., Donsker, D. & Rasmussen, P. (Eds). 2021. ***IOC World Bird List*** (v11.2). https://www.worldbirdnames.org

Jeanne Melchels

Jeanne is an independent illustrator, graphic designer and 2D motion designer from The Netherlands, who works from a shared studio space in an atmospheric 1940s high school building located in Arnhem. She is educated in both design and marketing communications. Being accurate and curious by nature, Jeanne specializes in distinctive visual communication, such as clear illustrations for complex messages or subjects. Most of her (Dutch) commissioned work consists of communication design: illustrations or infographics for governmental organizations, corporate identities for businesses or book covers, and educational illustrations for publishers. Nature's shapes and colors—such as hummingbirds with their dazzling plumage and iridescent feathers—are an endless source of inspiration to Jeanne. In 2016 she started a personal project on Instagram in which she is planning to illustrate 300 different hummingbird species in different styles, materials or techniques, a selection of which is shown *here*. It was through Instagram that Jeanne first came in contact with Glenn Bartley, having admired his fantastic hummingbird photos. Jeanne feels truly honored to be able to contribute to this stunning new book and, through her artwork, provide a different perspective on the wonderful world of hummingbirds.

To see more of Jeanne's work please visit: *Hummingbird illustration project*: https://www.instagram.com/jeanne_melchels; *Websites*: https://www.hummingbird-illustrations.com; https://www.jeannedesign.nl (in Dutch)

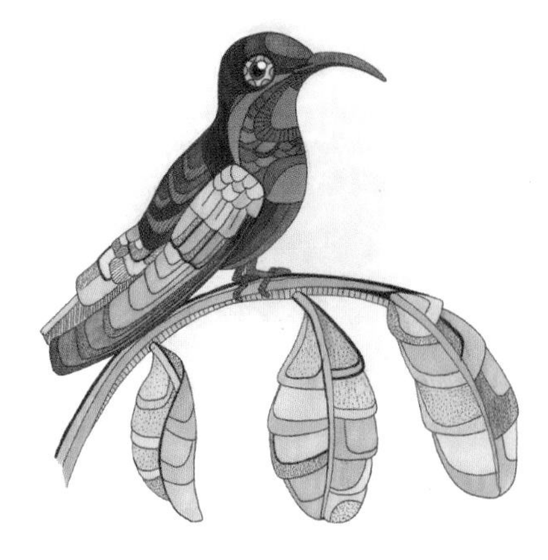

Acknowledgements and Photo Credits

The production of this book has been a collaborative effort involving many people, to all of whom we extend our grateful thanks. We would, however, particularly like to thank **Rob Hume** and **Chris Sharpe** for their invaluable help in researching and drafting elements of the text, and **Jeanne Melchels** for providing the wonderful illustrations that grace many of the pages. A brief biography of each of these key contributors is included here. Special thanks are also due to **Gill Swash** for her endless patience and painstaking work in editing and honing the draft text, and to the many photographers who contributed willingly to the project (see *next page*).

The helpful and enthusiastic contributions from the following friends and colleagues are also gratefully acknowledged: **Mark Balman** (BirdLife International), for compiling the hummingbird 'heat map' that appears on *page 115* and producing the species maps that appear in the section *The BirdLife List of Species*; **Ian Burfield** (BirdLife International), for his unwavering help in coordinating BirdLife's contribution to the book; **David Christie**, for his patience and keen eye for detail when copy editing the text; **Andy King** (principal geologist at environmental consultancy Geckoella), for his input to the section on *The origin of hummingbirds* on *page 11*; **Robert Kirk** (Publisher, Princeton Nature), for his support and guidance throughout the project; **Jessica Law**, **Jim Lawrence** and **Christopher Sands** (all BirdLife International), for their help, guidance and input in various ways; **Robert Still** (Princeton **WILD***Guides'* publishing director and chief graphic designer), for weaving his magic in presenting the material in the book to such great effect, and his contribution to the chapter *Color and Iridescence: Flashes of Brilliance*; and **Patricia Zurita** (BirdLife International), for kindly contributing the *Foreword*.

Thanks are also due to a number of other people who made invaluable contributions at various stages during the production of the book: **Stella Clifford-Jones**, **Nigel Collar**, **Paul Donald**, **Robert Edgar**, **Lance Grande**, **Katie Hoff**, **Steve Holmes**, **Martin Jones**, **Daniel Ksepka**, **David and Nancy Massie**, **Sarah McKenzie**, **Peter Mullen**, **Georg Pohland**, **Debby Reynolds**, **Bob Self** and **Nick Wilcox-Brown**.

Rob Hume

Rob is a writer and editor with a lifelong enthusiasm for birds. Further education in art and geography influence the way he appreciates the natural world. He began survey work for the Royal Society for the Protection of Birds (RSPB) in Wales, then moved to its Bedfordshire HQ. While helping to organize local groups and introduce celebrity speakers and RSPB film shows, he began editorial work for the journal *British Birds* and later became a member of its editorial board and Chairman of its Rarities Committee. He transferred to writing and editing RSPB publications, from books and magazines to annual reviews; under his editorship, both *Bird Life* and *Birds* magazines won national media awards. Many books followed, including collaborations with artist Peter Hayman and the **WILD***Guides* team to produce best-selling identification guides. Widely travelled, he often cites, as one of his favorite places, the beaches of Tobago, with their unbeatable combination of magnificent frigatebirds and sensational hummingbirds.

Rob made significant contributions to the following chapters: *An Introduction to Hummingbirds* (*page 9*), *Adaptations for Exceptional Lifestyles* (*page 15*), *Color and Iridescence: Flashes of Brilliance* (*page 49*), *Breeding: Continuing the Line* (*page 87*), *Biogeography and Biodiversity: The Hummingbirds' Realm* (*page 115*), and *Hummingbirds and People: History, Discovery and Culture* (*page 179*).

Christopher J. Sharpe

Chris has worked on the conservation of Neotropical birds since the 1980s, having been based for most of that time in Venezuela, where he is a Research Associate at the *Phelps Ornithological Collection* (*COP*) and the NGO *Provita*, a Founder Member of the *Venezuelan Ornithologists' Union* (sitting on the *Venezuelan Bird Records Committee*), and Birds Editor for the IUCN *Red Data Book of Venezuelan Fauna*. Formerly an editor of *HBW Alive* and the *Lynx and BirdLife International Field Guides series*, he is currently Associate Editor of *Birds of the World* at the *Cornell Lab of Ornithology* and Associate Editor of the *Bulletin of the British Ornithologists' Club*. His bird identification texts have appeared in several field guides and he is a co-author (with Guy Kirwan) of the standard *Birds of the West Indies*. Chris works as a consultant on Latin American conservation for a range of international organizations and governments. Over the course of his career, he has been fortunate to work throughout the Americas from Alaska to Tierra del Fuego, and to lead over a hundred bird tours, providing the ideal opportunity to enjoy and share encounters with so many of the wonderful hummingbirds featured in this book.

Chris made significant contributions to the following chapters: *Conservation: Hummingbirds Under Threat* (*page 191*), and *Taxonomy: The BirdLife List of Species* (*page 239*).

Photo credits

Our aspiration in producing this book has been to compile an unsurpassed collection of high-quality images that show all aspects of hummingbird biology and behavior, and to illustrate as wide a range of species as possible. Although the majority (73%) of the 547 photos that are featured were taken by us, we are greatly indebted to the many wildlife photographers from around the world who kindly responded to our requests for images and enabled us to fill the inevitable gaps. In total, 80 photographers have contributed images: all are listed below with details of their website (where appropriate) and the number of images that are featured [in square brackets].

Roger Ahlman [1]; **Ciro Albano** (brazilbirdingexperts.com) [17]; **Fabio Arias** (flickr.com/photos/133729470@N06) [1]; **Enrique Ascanio** [1]; **Glenn Bartley** (glennbartley.com) [316]; **Greg Basco** (deepgreenphotography.com) [3]; **Simon Best** [1]; **Ads Bowley** [1]; **Dušan Brinkhuizen** (sapayoa.com) [1]; **Caio Brito** (brazilbirdingexperts.com) [1]; **Carlos Calle** (flickr.com/photos/guiacalles) [4]; **José Cañas** (flickr.com/photos/jose_canas) [3]; **Angel Cárdenas Hidalgo** [1]; **Victor Castanho** (instagram.com/victorbirdphotography) [1]; **Pablo Cerqueira** (brazilbirdingexperts.com) [2]; **Dustin Chen** (flickr.com/photos/27424285@N00) [6]; **Oswaldo Cortes and Daira Ximena Villagran** (flickr.com/photos/oswaldo_cortes_aves) [1]; **Jan Axel Cubilla Rodríguez** [2]; **Rob Curtis** (theearlybirder.org) [1]; **Greg and Yvonne Dean** (agami.nl) [4]; **Knut Eisermann** (knut-eisermann.info) [3]; **Jess Findlay** (jessfindlay.com) [2, including scenery on *page 136*]; **Tom Friedel** (BirdPhotos.com) [2]; **Jacob Garvelink** (agami.nl) [1]; **Jaime Andres Herrera Villarreal** (flickr.com/photos/herrinsa) [2]; **Greg Homel** (naturalencountersbirdingtours.com) [2, including scenery on *page 128*]; **Serrafin Robert Horst** [1]; **Dave Irving** (flickr.com/photos/dave_irving) [1]; **Daniel Ksepka** [1]; **Felicia Kulp/UT Austin** [1]; **Jim Lawrence** [1]; **Daniel López-Velasco** (agami.nl) [3]; **Pio Marshall** [1]; **Eric Antonio Martinez** [1]; **David Massie** [1 (coastal desert, *page 135*)]; **Amy McAndrews** (flickr.com/photos/34470420@N02) [1]; **Robin Mettler** [3]; **Denzil Morgan** (facebook.com/Denzil-Morgan-Photography-786259851389520) [8]; **Pete Morris** (agami.nl) [3]; **Alan Murphy** (alanmurphyphotography.com) [5]; **Jonathan Newman** [1]; **Brandon Niddifer** [1]; **Phil Palmer** [1]; **José M. Pantaleón** [1]; **Bruno Rennó** (brazilbirdingexperts.com) [1]; **Manuel Roncal-Rabanal** [1]; **Sebastian Saiter** [1]; **Norton Santos** (flickr.com/photos/nortondefeis) [2]; **Dubi Shapiro** (pbase.com/dubisha) [6]; **Charles Sharp** (flickr.com/photos/93882360@N07) [1]; **Christopher J. Sharpe** [1 (scenery, *page 123*)]; **Jean Simard** (flickr.com/photos/jeannot7) [1]; **Petr Simon** [1]; **Brian E. Small** (briansmallphoto.com) [2]; **David J. Southall** (tropicalbirdphotos.com) [2]; **Jacob S. Spendelow** (tringa.org) [1]; **Matthew Studebaker** (studebakerstudio.com) [2]; **Robert Still** [1 (soap bubble, *page 66*)]; **Andy and Gill Swash** (agami.nl) [84]; **Vivek Tiwari** (flickr.com/photos/spiderhunters) [1]; **Juan Diego Vargas** (birdingwithjuandiego.com) [2]; **Andrés Vasquez Noboa** [2]; **Frédéric Weckstéen** (flickr.com/photos/147450357@N08) [1]; **Tiny and Jacob Wijpkema** [1] and **Tim Zurowski** (timzurowski.smugmug.com) [2].

In addition, ten images are reproduced under the terms of the following Creative Commons licenses: Attribution 2.0 Generic (CC BY 2.0), Attribution-ShareAlike 2.0 Generic (CC BY-SA 2.0) or Attribution-ShareAlike 4.0 International (CC BY-SA 4.0). These are: *page 80* Rainbow **Mavrica** (CC BY-SA 2.0); *page 135* Temperate rainforest **Hector Montero** (CC BY-SA 2.0); *page 158* Scenery **Douglas Scortegagna** (CC BY 2.0); *page 174* Sapsucker holes **Under the same moon** (CC BY 2.0), Jewelweed **Alvin Kho** (CC BY 2.0), Turk's Cap **Forest and Kim Starr** (CC BY 2.0), Dwarf Buckeye **De Tuin** (Public Domain), Trumpet Creeper **Carl Lewis** (CC BY-SA 2.0); *page 179* Moctezuma's feather headdress **Thomas Ledl** (CC BY-SA 4.0); *page 180* Nazca lines **Diego Delso** (CC BY-SA 4.0).

A 'thumbnail' image of each of the hummingbird photos provided by the contributing photographers, except for those taken by the authors, is shown *here*.

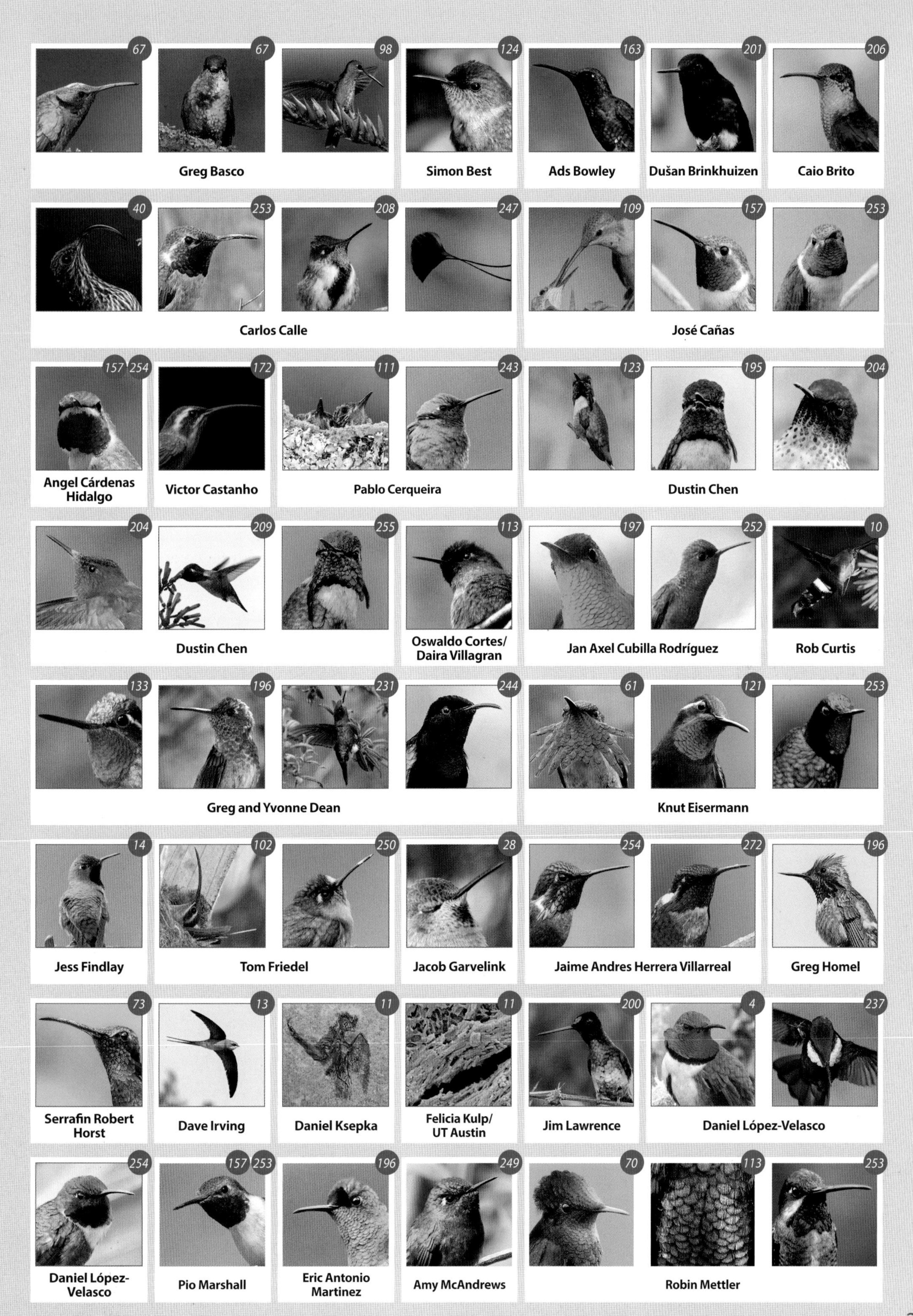
67
67
98
124
163
201
206
Greg Basco
Simon Best
Ads Bowley
Dušan Brinkhuizen
Caio Brito
40
253
208
247
109
157
253
Carlos Calle
José Cañas
157 254
172
111
243
123
195
204
Angel Cárdenas Hidalgo
Victor Castanho
Pablo Cerqueira
Dustin Chen
204
209
255
113
197
252
10
Dustin Chen
Oswaldo Cortes/ Daira Villagran
Jan Axel Cubilla Rodríguez
Rob Curtis
133
196
231
244
61
121
253
Greg and Yvonne Dean
Knut Eisermann
14
102
250
28
254
272
196
Jess Findlay
Tom Friedel
Jacob Garvelink
Jaime Andres Herrera Villarreal
Greg Homel
73
13
11
11
200
4
237
Serrafin Robert Horst
Dave Irving
Daniel Ksepka
Felicia Kulp/ UT Austin
Jim Lawrence
Daniel López-Velasco
254
157 253
196
249
70
113
253
Daniel López-Velasco
Pio Marshall
Eric Antonio Martinez
Amy McAndrews
Robin Mettler

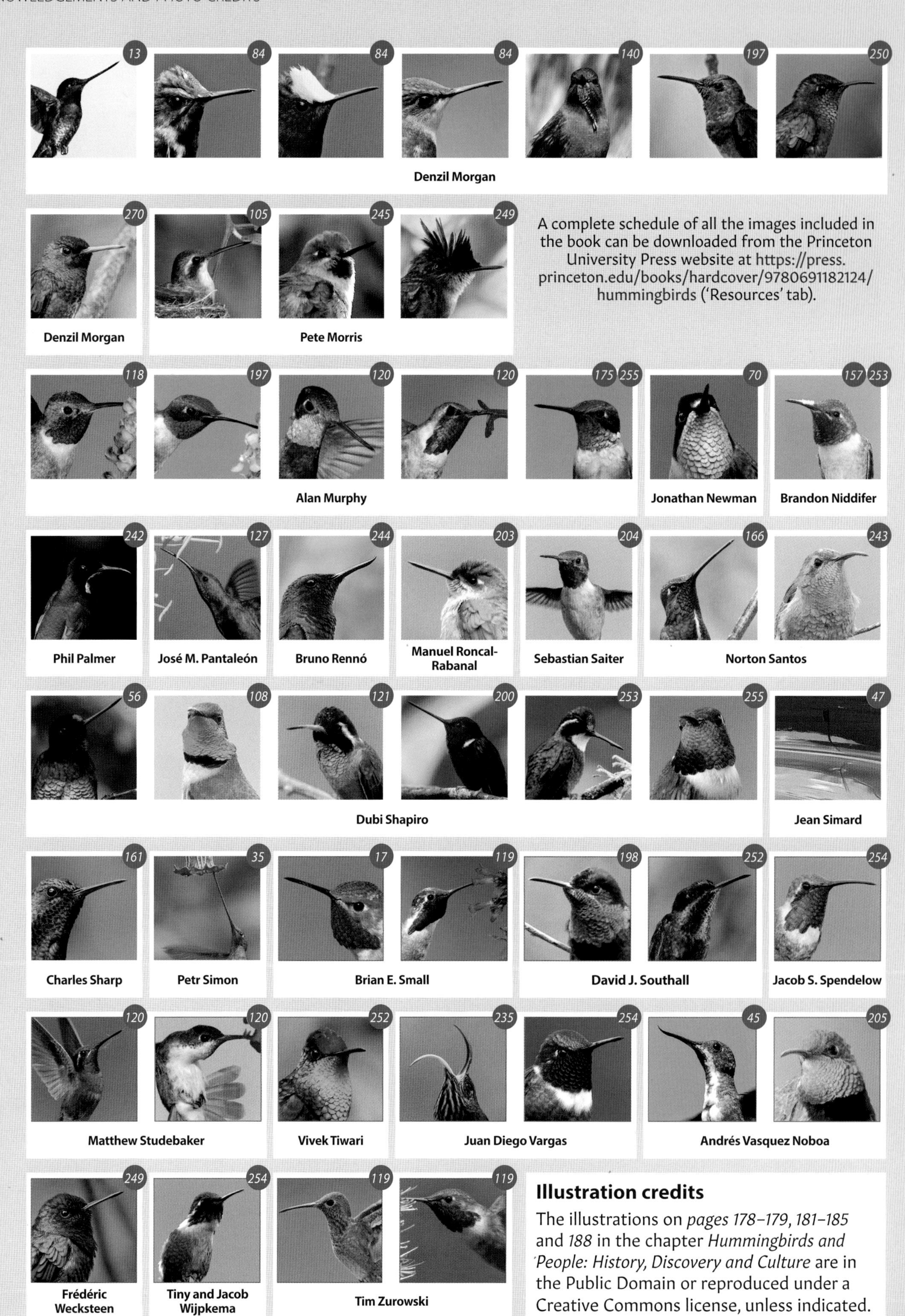

A complete schedule of all the images included in the book can be downloaded from the Princeton University Press website at https://press.princeton.edu/books/hardcover/9780691182124/hummingbirds ('Resources' tab).

Illustration credits

The illustrations on *pages 178–179, 181–185* and *188* in the chapter *Hummingbirds and People: History, Discovery and Culture* are in the Public Domain or reproduced under a Creative Commons license, unless indicated.

Index

This index includes the English and, in *italics, scientific* names of all the hummingbirds covered in this book. The English names of species that are illustrated with a photograph are highlighted in **bold** text.

Alternative English names used by taxonomic authorities other than BirdLife International are shown in grey text.

Bold italicized numbers refer to a page where a photograph appears; regular *italicized* numbers indicate illustrations; blue numbers are used for the genus (**bold**) and species entries in *The BirdLife List of Species*. Regular figures refer to pages where other key information can be found.

A

Abeillia 249, 266
— *abeillei* ***249***, 266
Adelomyia 244, 260
— *melanogenys* ***104***, ***151***, ***244***, 260
Aglaeactis 247, 263
— *aliciae* 263
— *castelnaudii* ***247***, 263
— *cupripennis* ***43***, ***49***, ***56***, ***105***, ***140***, 263
— *pamela* ***224***, 263
Aglaiocercus 245, 260
— *berlepschi* 260
— *coelestis* ***86***, ***147***, 260
— *kingii* ***18***, *20*, ***61***, ***113***, ***228***, ***245***, 260, 275, 276
Amazilia 106, 251, 268
— *alfaroana* 269
— *amabilis* ***8***, 269, 275
— *amazilia* ***156***, 268
— *bartletti* 269
— *beryllina* 269
— *boucardi* ***197***, 269
— *brevirostris* 268
— *candida* 268
— *castaneiventris* ***200***, 268
— *chionogaster* ***268***, 268, 275
— *chrysura* 270
— *coeruleogularis* ***70***, 270
— *cupreicauda* 269
— *cyanifrons* ***269***, 269
— *cyanocephala* 269
— *cyanura* 269
— *cyanus* ***29***, 270, 276
— *decora* 269
— *edward* ***88***, 269
— *eliciae* ***52***, 270
— *fimbriata* ***30***, ***154***, 269, 275, 276
— *franciae* ***25***, ***148***, 268, 275
— *goudoti* ***270***, 270
— *grayi* 270
— *humboldtii* ***270***, 270
— *julie* ***36***, ***47***, ***76***, 270
— *lactea* ***51***, 269
— *leucogaster* ***163***, 268
Amazilia lilliae ***190***, ***199***, 270
— *luciae* ***196***, 269
— *oenone* ***9***, ***155***, 270
— *rosenbergi* ***269***, 269
— *rutila* ***130***, 268
— *sapphirina* ***251***, 270
— *saucerottei* ***32***, 269
— *tobaci* ***162***, 270
— *tzacatl* ***101***, ***111***, ***268***, 268
— *versicolor* ***171***, 268
— *violiceps* ***120***, 269
— *viridicauda* 268
— *viridifrons* 269
— *viridigaster* 269
— *wagneri* 269
— *yucatanensis* ***119***, 268
Androdon 243, 258
— *aequatorialis* 39, ***45***, ***243***, 258
Anopetia 242, 256
— *gounellei* ***242***, 256
Anthocephala 250, 266
— *berlepschi* ***250***, 266
— *floriceps* ***138***, 266, 267
Anthracothorax 244, 258
— *aurulentus* 259
— *dominicus* ***127***, 259
— *mango* ***124***, ***259***, 259
— *nigricollis* ***161***, ***244***, 258
— *prevostii* ***21***, ***121***, ***130***, 258
— *veraguensis* 259
— *viridigula* 258
— *viridis* 259
Aphantochroa 250, 266
— *cirrochloris* ***171***, ***250***, 266
Archilochus 255, 273
— *alexandri* ***110***, ***118***, 273
— *colubris* ***118***, ***175***, ***211***, ***229***, ***255***, 273
Atthis 255, 273
— *ellioti* ***60***, 273
— *heloisa* ***89***, ***255***, 273
Augastes 243, 258
— *lumachella* ***96***, ***167***, 258
— *scutatus* ***167***, ***243***, 258
Avocetbill, Mountain ***245***, 261
Avocettula 244, 258
— *recurvirostris* 39, ***45***, ***244***, 258
Awlbill, Fiery-tailed 39, ***45***, ***244***, 258

B

Barbthroat, Band-tailed 90, ***242***, 256
—, **Pale-tailed** 90, ***165***, 256, 274
—, Sooty 256
Basilinna 252, 270
— *leucotis* ***108***, ***121***, ***252***, 270
— *xantusii* ***121***, 270
Blossomcrown 267
—, **Santa Marta** ***138***, 199, 266
—, **Tolima** 199, ***250***, 266
Boissonneaua 248, 264
— *flavescens* ***94***, ***151***, ***248***, 264
— *jardini* ***23***, ***30***, ***42***, ***57***, ***71***, ***148***, 264
— *matthewsii* ***44***, ***99***, ***152***, ***189***, 264
Brilliant, Black-breasted 241, 264, 265
—, **Black-throated** ***241***, 264
—, **Empress** ***150***, 265
—, **Fawn-breasted** ***151***, 265
—, **Gould's** ***165***, 265
—, **Green-crowned** ***42***, ***47***, ***82***, 90, ***98***, ***222***, 265
—, Pink-throated 199, 201, 264
—, Rufous-webbed 264
—, Velvet-browed 264
—, **Violet-fronted** ***248***, 265

C

Calliphlox 254, 272
— *amethystina* ***100***, ***104***, ***161***, 272, 275, 276
— *bryantae* 272
— *mitchellii* ***74***, ***254***, 272
Calothorax 254, 273
— *lucifer* ***120***, ***254***, 273
— *pulcher* 273
Calypte 106, 255, 273
— *anna* ***28***, ***31***, ***38***, 44, ***47***, 52, ***69***, 92, 106, ***118***, ***210***, ***255***, 273
— *costae* 39, ***119***, 273

Campylopterus 250, 266
— *calcirupicola* ***239***, 266, 267
— *curvipennis* 266, 267
— *cuvierii* ***266***, 266
— *diamantinensis* 266, 267
— *duidae* 267
— *ensipennis* ***163***, 267
— *excellens* 267
— *falcatus* ***250***, 267
— *hemileucurus* ***129***, ***221***, 267
— *hyperythrus* 266
— *largipennis* ***165***, 266, 267
— *phainopeplus* 267
— *rufus* ***266***, 266
— *villaviscensio* ***22***, 267
Carib, Green-throated ***82***, 122, ***126***, 259
—, **Purple-throated** *39*, ***85***, 122, ***126***, ***244***, 259
Chaetocercus 254, 272
— *astreans* 272
— *berlepschi* ***201***, 272
— *bombus* 272
— *heliodor* ***254***, ***272***, 272, 276
— *jourdanii* 272
— *mulsant* ***13***, ***22***, ***150***, 272, 276
Chalcostigma 246, 261
— *herrani* ***81***, ***101***, 261, 275
— *heteropogon* 261
— *olivaceum* ***140***, 261
— *ruficeps* ***246***, 261
— *stanleyi* ***261***, 261
Chalybura 250, 267
— *buffonii* ***250***, 267
— *urochrysia* ***267***, 267
Chionomesa viridifrons 269
Chlorostilbon 249, 265
— *alice* 265, 275
— *assimilis* ***13***, 265
— *auriceps* 265
— *bracei* 194
— *canivetii* 265
— *elegans* 194
— *forficatus* 265
— *gibsoni* ***105***, 265
— *lucidus* ***249***, 265, 276
— *maugaeus* 265
— *melanorhynchus* 265
— *mellisugus* ***265***, 265
— *notatus* ***161***, 265, 275, 276
— *olivaresi* 265
— *poortmani* 265, 275, 276
— *ricordii* ***123***, 265
— *russatus* 265
— *stenurus* 265
— *swainsonii* 265
Chrysolampis 243, 258
— *mosquitus* ***70***, ***95***, ***110***, ***237***, ***177***, ***243***, 258, 275
Clytolaema 248, 265
— *rubricauda* ***23***, ***30***, ***50***, ***170***, ***248***, 265
Coeligena 247, 263
— *albicaudata* 263
— *bonapartei* ***9***, ***13***, ***113***, 264
— *coeligena* ***247***, 263, 276
— *conradii* 263
— *consita* 264
— *dichroura* 263
— *eisenmanni* 263
— *eos* 264
— *helianthea* ***54***, ***113***, ***213***, 264
— *inca* 263
— *iris* ***54***, ***217***, 263
— *lutetiae* ***27***, ***64***, 264, 276
— *orina* ***56***, ***200***, 264
— *osculans* ***263***, 263
— *phalerata* ***79***, ***99***, 264
— *prunellei* ***200***, 263, 275, 276
— *torquata* ***37***, ***59***, 263, 276
— *violifer* 263
— *wilsoni* ***32***, ***146***, 263
Colibri 106, 243, 258
— *coruscans* ***99***, ***111***, ***141***, 258, 274
— *cyanotus* 259
— *delphinae* ***95***, ***146***, ***258***, 258
— *serrirostris* ***243***, 258
— *thalassinus* ***67***, ***238***, 258, 259
Comet, Bronze-tailed ***245***, 261
—, **Grey-bellied** ***203***, ***245***, 260
—, **Red-tailed** *20*, ***48***, ***159***, ***245***, 260
Coquette, Black-crested ***83***, ***91***, ***129***, 260
—, Butterfly 260, 261, 275
—, Dot-eared 205, 206, 260
—, **Festive** ***29***, ***76***, ***171***, 205, 260, 261
—, **Frilled** ***169***, ***236***, 260
—, Peacock 260
—, **Racket-tailed** ***96***, 260
—, **Rufous-crested** ***61***, ***97***, ***260***, 260, 274
—, **Short-crested** 195, ***196***, 260
—, **Spangled** ***244***, 260, 275
—, **Tufted** ***6***, ***10***, ***91***, 162, 260
—, **White-crested** ***41***, 260
Coronet, Buff-tailed ***94***, ***151***, ***248***, 264
—, **Chestnut-breasted** ***44***, ***99***, ***152***, ***189***, 264
—, **Velvet-purple** ***23***, ***30***, ***42***, ***57***, ***71***, ***148***, 264
Cyanophaia 249, 266
— *bicolor* ***249***, 266
Cynanthus 249, 266
— *doubledayi* 266, 267
— *latirostris* ***24***, ***119***, ***249***, 266, 267
— *lawrencei* 266, 267
— *sordidus* 266

D
Daggerbill, Geoffroy's 137
—, White-throated 137
Discosura 244, 260
— *conversii* ***47***, ***149***, ***244***, 260
— *langsdorffi* 260
— *letitiae* 260
— *longicaudus* ***96***, 260
— *popelairii* ***93***, ***202***, 260
Doricha 254, 273
— *eliza* ***133***, 273
— *enicura* ***254***, 273
Doryfera 243, 258
— *johannae* 258
— *ludovicae* *39*, ***151***, ***243***, 258

E
Elvira 250, 267
— *chionura* 267
— *cupreiceps* ***132***, ***250***, 267
Emerald, Andean ***25***, ***148***, 268, 275
—, **Blue-chinned** ***161***, 265, 275, 276
—, **Blue-tailed** ***265***, 265
—, Brace's 194
—, Canivet's 265
—, Caribbean 194
—, Chiribiquete 265
—, Coppery 265
—, **Coppery-headed** 90, ***132***, ***250***, 267
—, Cozumel 265
—, **Cuban** 122, ***123***, 188, 265
—, **Garden** ***13***, 265
—, **Glittering-bellied** ***249***, 265, 276
—, **Glittering-throated** ***30***, ***154***, 269, 275, 276
—, Golden-crowned 265
—, Green-tailed 265, 276
—, Hispaniolan 122, 265
—, **Honduran** ***196***, 269
—, Narrow-tailed 265
—, **Plain-bellied** ***163***, 268
—, Puerto Rican 122, 265
—, **Red-billed** ***105***, 265
—, **Sapphire-spangled** ***51***, 269
—, Short-tailed 265, 275, 276
—, Spot-vented 269
—, **Versicolored** 168, ***171***, 268
—, Western 265
—, White-bellied 90, 268
—, White-chested 268
—, White-tailed 90, 267
Ensifera 247, 264
— *ensifera* ***35***, *39*, ***144***, ***247***, 264
Eriocnemis 246, 262
— *aline* ***50***, 263
— *cupreoventris* ***52***, 262, 274, 275, 276
— *derbyi* ***262***, 262
— *glaucopoides* 263
— *godini* 262

Eriocnemis *isabellae* 262
— *luciani* ***219***, 262, 263, 276
— *mirabilis* 263
— *mosquera* ***52***, ***246***, 262
— *nigrivestis* ***201***, ***231***, 262, 276
— *sapphiropygia* 262, 263
— *vestita* ***76***, 262, 274, 276
Eugenes 252, 271
— *fulgens* ***56***, ***70***, ***120***, ***252***, 270, 271
— *spectabilis* 270
Eulampis 244, 259
— *holosericeus* ***82***, ***126***, 259
— *jugularis* *39*, ***85***, ***126***, ***244***, 259
Eulidia 253, 272
— *yarrellii* ***157***, ***204***, ***253***, 272
Eupetomena 250, 267
— *macroura* *20*, ***160***, ***250***, 267
Eupherusa 250, 267
— *cyanophrys* ***196***, 267
— *eximia* ***130***, ***250***, 267
— *nigriventris* ***25***, ***59***, ***267***, 267
— *poliocerca* 267
Eutoxeres 242, 256
— *aquila* ***40***, ***235***, ***242***, 256
— *condamini* *39*, 256

F

Fairy, Black-eared 258
—, **Purple-crowned** ***87***, *183*, ***220***, ***243***, 258
Firecrown, **Green-backed** ***158***, 174, 204, ***244***, 259
—, **Juan Fernandez** 158, ***204***, 259
Florisuga 242, 256
— *fusca* ***24***, ***79***, ***169***, ***256***, 256
— *mellivora* ***13***, ***79***, ***154***, ***242***, 256
Florisuginae 12, 13, 240, 242, 256

G

Glaucis 106, 242, 256
— *aeneus* *39*, 256
— *dohrnii* ***103***, ***205***, 256
— *hirsutus* ***242***, 256, 274
Goethalsia 252, 270
— *bella* ***197***, ***252***, 270
Goldentail, **Blue-throated** ***52***, 90, 270
Goldenthroat, **Green-tailed** ***111***, ***163***, 258
—, Tepui 258
—, **White-tailed** ***243***, 258
Goldmania 252, 270
— *violiceps* ***252***, 270

H

Haplophaedia 246, 262
— *assimilis* 262
— *aureliae* ***246***, 262
— *lugens* ***148***, 262
Heliactin 243, 258
— *bilophus* ***13***, ***110***, ***167***, ***243***, 258
Heliangelus 244, 259
— *amethysticollis* ***50***, ***244***, 259, 274, 275
— *clarisse* 259, 274
— *exortis* ***42***, ***105***, ***152***, 259
— *mavors* ***64***, 259
— *micraster* ***259***, 259
— *regalis* ***74***, ***203***, 259
— *spencei* 259
— *strophianus* ***74***, 259
— *viola* ***54***, ***217***, 259
— *zusii* 112, 275
Heliodoxa 248, 264
— *aurescens* ***165***, 265
— *branickii* 264
— *gularis* 264
— *imperatrix* ***150***, 265
— *jacula* ***42***, ***47***, ***82***, ***98***, ***222***, 265
— *leadbeateri* ***248***, 265
— *rubinoides* ***151***, 265
— *schreibersii* ***241***, 264
— *whitelyana* 264, 265
— *xanthogonys* 264
Heliomaster 253, 271
— *constantii* 271
— *furcifer* ***73***, 271
— *longirostris* ***253***, ***271***, 271
— *squamosus* ***166***, *183*, ***227***, 271
Heliothryx 243, 258
— *auritus* 258
— *barroti* ***87***, *183*, ***220***, ***243***, 258
Helmetcrest, Bearded 261
—, Blue-bearded 138, 199, 261
—, **Buffy** 199, ***224***, 261
—, **Green-bearded** ***140***, ***246***, 261
—, White-bearded 261
Hermit, Ash-bellied 257
—, Black-throated 256
—, **Broad-tipped** ***242***, 274, 256
—, Bronzy *39*, 256
—, Buff-bellied 257
—, **Cinnamon-throated** 256, ***274***
—, **Dusky-throated** ***172***, 256
—, **Ecuadorian** ***9***, 257
—, **Great-billed** ***89***, 90, ***102***, 257
—, **Green** ***74***, 87, 90, ***221***, 257
—, **Grey-chinned** ***13***, 90, 257, 274
—, **Hook-billed** ***103***, 168, ***205***, 256
—, **Koepcke's** ***155***, 203, 257
—, **Little** 90, 162, ***163***, 256
—, **Long-billed** ***33***, ***131***, 257
—, **Long-tailed** ***60***, 90, 257
—, Mexican 257
—, **Minute** 90, ***172***, 256
—, Needle-billed 257
—, Pale-bellied 257
Hermit, **Planalto** ***257***, 257
—, Porculla 257
—, **Reddish** ***41***, 78, ***89***, 90, ***103***, 257
—, **Rufous-breasted** 87, 122, ***242***, 256, 274
—, **Saw-billed** ***45***, ***173***, ***242***, 256
—, **Scale-throated** ***242***, 257, 274
—, Sooty-capped 90, 101, 257, 274
—, Straight-billed 90, 257
—, **Streak-throated** ***89***, 256
—, Stripe-throated 257
—, Tapajos 205, 206, 256
—, Tawny-bellied 90, 257
—, White-bearded 90, 257
—, White-browed 257
—, **White-whiskered** 90, ***155***, 257
Hillstar, **Andean** ***108***, 261
—, **Black-breasted** ***207***, 261
—, **Blue-throated** ***4***, 5, 201, ***207***, 241, 261
—, **Ecuadorian** ***55***, 207, ***218***, ***245***, 261
—, Green-headed 207, 261
—, **Rufous-gaped** ***146***, 264
—, **Wedge-tailed** ***205***, 261
—, **White-sided** ***158***, 205, 261
—, **White-tailed** ***248***, 264
Hummingbird, **Allen's** ***17***, 92, ***119***, 273
—, **Amazilia** ***156***, 268
—, **Amethyst-throated** 90, 117, ***121***, 271
—, **Anna's** ***28***, ***31***, ***38***, *44*, ***47***, *52*, 58, *68*, ***69***, 80, 90, *92*, 101, *106*, 117, ***118***, 176, 183, 195, ***210***, ***255***, 273
—, **Antillean Crested** *39*, ***95***, 122, ***127***, ***249***, 266
—, Azure-crowned 269
—, **Bahama** 117, 122, ***255***, 273
—, Beautiful 273
—, **Bee** *15*, *16*, 122, ***123***, 188, ***195***, ***209***, ***255***, 273
—, Berylline 117, 121, 269
—, **Black-bellied** ***25***, ***59***, 90, ***267***, 267
—, **Black-chinned** 108, ***110***, 117, ***118***, 195, 273
—, **Blue-chested** ***8***, 90, 269, 275
—, **Blue-headed** 122, ***249***, 266
—, Blue-tailed 269
—, **Blue-throated** ***120***, 271
—, Blue-vented 269
—, **Broad-billed** ***24***, ***119***, ***249***, 266, 267
—, **Broad-tailed** 62, 90, 108, ***118***, 273
—, **Buff-bellied** 108, ***119***, 268
—, **Buffy** ***162***, 268
—, **Bumblebee** ***89***, 117, 121, ***255***, 273
—, **Calliope** 90, 92, 108, 112, 117, ***119***, ***272***, 273
—, Charming 90, 269

Hummingbird, Chestnut-bellied 199, ***200***, 268
—, **Cinnamon** 90, 117, 121, ***130***, 268
—, Cinnamon-sided 269
—, **Copper-rumped** ***162***, 270
—, Copper-tailed 269
—, **Costa's** 39, 106, 112, 117, ***119***, 273
—, Doubleday's 267
—, Dusky 266
—, **Eastern Wedge-billed** ***137***, 258
—, **Emerald-chinned** ***249***, 266
—, **Fiery-throated** ***1***, ***67***, ***133***, ***252***, 271
—, **Garnet-throated** ***253***, 271
—, **Giant** 12, ***13***, 15, ***16***, 101, 158, ***177***, ***232***, ***249***, 265
—, Gilded 270
—, Glow-throated 197, 198, 273
—, Gray's 270
—, Green-and-white 268
—, Green-bellied 269
—, Green-fronted 269
—, Guanacaste 197, 241, 269
—, **Humboldt's** ***270***, 270
—, **Indigo-capped** 197, ***269***, 269
—, **Lucifer** ***120***, ***254***, 273
—, Lyre-tailed 122, 273
—, **Magnificent** ***56***, ***70***, 108, ***120***, ***238***, ***252***, 270, 271
—, **Mangrove** 128, ***197***, 269
—, **Many-spotted** ***251***, 268
—, **Oasis** 93, ***109***, ***157***, ***253***, 272
—, **Oaxaca** 195, ***196***, 267
—, Olive-spotted 268
—, **Pirre** ***197***, 199, ***252***, 270
—, **Purple-chested** ***269***, 269
—, Rivoli's 270
—, **Ruby-throated** 44, 98, 117, ***118***, 122, 174, ***175***, 176, 195, ***211***, ***229***, ***255***, 273
—, **Ruby-topaz** ***70***, ***95***, ***110***, 122, 176, ***177***, 187, ***237***, ***243***, 258, 275
—, **Rufous** 5, 90, 92, 101, 108, 117, ***119***, ***175***, 176, ***195***, 273
—, **Rufous-tailed** 90, 100, ***101***, ***111***, ***268***, 268
—, **Rufous-throated** ***251***, 270
—, **Sapphire-bellied** ***190***, ***199***, 270
—, **Sapphire-throated** ***70***, 270
—, **Scintillant** ***255***, 273
—, **Scissor-tailed** ***198***, ***252***, 271
—, **Shining-green** ***270***, 270
—, **Snowy-bellied** ***88***, 269
—, **Sombre** ***171***, ***250***, 266
—, **Speckled** ***104***, ***151***, ***244***, 260
—, Spot-throated 268
—, **Steely-vented** ***32***, 197, 269
—, **Stripe-tailed** 128, ***130***, ***250***, 267
—, **Swallow-tailed** 20, 90, ***160***, ***250***, 267
Hummingbird, Sword-billed 29, ***35***, 39, ***144***, ***247***, 264
—, Talamanca 270
—, **Tooth-billed** 39, 44, ***45***, 90, ***243***, 258
—, Tres Marias 195, 266, 267
—, **Tumbes** ***156***, ***251***, 268
—, Turquoise-crowned 266, 267
—, **Vervain** 106, 122, ***124***, 273
—, **Violet-bellied** ***36***, ***47***, ***76***, 270
—, **Violet-capped** 197, 199, ***252***, 270
—, **Violet-chested** ***252***, 271
—, **Violet-crowned** 117, ***120***, 269
—, **Violet-headed** 90, ***155***, ***249***, 266
—, **Volcano** ***273***, 273
—, Wedge-billed 34, 259
—, **Western Wedge-billed** ***39***, ***137***, ***243***, 258
—, **White-bellied** ***268***, 268, 275
—, **White-eared** 90, ***108***, ***121***, ***252***, 270
—, White-tailed 195, 267
—, **White-throated** ***251***, 268, 276
—, **Wine-throated** ***60***, 273
—, **Xantus's** 117, ***121***, 270
Hylonympha 252, 271
— *macrocerca* ***198***, ***252***, 271

I
Inca, Black 199, ***200***, 263, 275, 276
—, **Bronzy** ***247***, 263, 276
—, **Brown** ***32***, ***146***, 263
—, **Collared** ***37***, ***59*** , 263, 276
—, Gould's 263
—, Green 263
—, Vilcabamba 263

J
Jacobin, Black ***24***, ***79***, 168, ***169***, ***256***, 256
—, **White-necked** ***13***, ***79***, 122, ***154***, ***242***, 256

K
Klais 249, 266
— *guimeti* ***155***, ***249***, 266

L
Lafresnaya 247, 264
— *lafresnayi* ***247***, 264, 276
Lampornis 253, 271
— *amethystinus* ***121***, 271
— *calolaemus* ***133***, 271
— *castaneoventris* ***50***, ***253***, 271
— *cinereicauda* ***83***, 271
— *clemenciae* ***120***, 271
— *hemileucus* ***13***, 271
— *sybillae* 271
— *viridipallens* ***133***, 271
Lamprolaima 253, 271
— *rhami* ***253***, 271
Lancebill, Blue-fronted 258
—, **Green-fronted** 39, ***151***, ***243***, 258
Lesbia 246, 261
— *nuna* ***246***, 261, 275
— *victoriae* ***14***, 261, 275, 276
Lesbiinae 12, 13, 240, 244–248, 259
Leucippus 251, 268
— *baeri* ***156***, ***251***, 268
— *chlorocercus* 268
— *fallax* ***162***, 268
— *taczanowskii* 268
Leucochloris 251, 268
— *albicollis* ***251***, 268, 276
Leucolia viridifrons 269
Loddigesia 247, 263
— *mirabilis* 20, ***27***, ***92***, 185, ***208***, ***223***, ***247***, 263
Lophornis 244, 260
— *adorabilis* ***41***, 260
— *brachylophus* ***196***, 260
— *chalybeus* ***29***, ***76***, ***171***, 260, 261
— *delattrei* ***61***, ***97***, ***260***, 260, 274
— *gouldii* 260
— *helenae* ***83***, ***91***, ***129***, 260
— *magnificus* ***169***, ***236***, 260
— *ornatus* ***6***, ***10***, ***91***, 260
— *pavoninus* 260
— *stictolophus* ***244***, 260, 275
— *verreauxii* 260, 261, 275

M
Mango, Antillean 259
—, **Black-throated** ***161***, ***244***, 258, 275
—, Green 122, 259
—, **Green-breasted** ***21***, 90, 117, ***121***, 122, 128, ***130***, 258
—, Green-throated 258
—, **Hispaniolan** 122, ***127***, 259
—, **Jamaican** 122, ***124***, ***259***, 259
—, Puerto Rican 122, 259
—, Veraguas 259
Mellisuga 255, 273
— *helenae* 15, 16, ***123***, ***195***, ***209***, ***255***, 273
— *minima* ***124***, 273
Metallura 246, 262
— *aeneocauda* ***262***, 262
— *baroni* ***201***, 262
— *eupogon* 262
— *iracunda* 262
— *odomae* 262
— *phoebe* ***28***, 262
— *theresiae* ***53***, 262, 274
— *tyrianthina* ***113***, ***143***, 262
— *williami* ***246***, 262
Metaltail, Black ***28***, 29, 262

Metaltail, Coppery ... ***53***, 262, 274
—, Fire-throated ... 262
—, Neblina ... 262
—, Perija ... 198, 199, 262
—, **Scaled** ... ***262***, 262
—, **Tyrian** ... ***113***, ***143***, 262
—, **Violet-throated** ... ***201***, 202, 262
—, **Viridian** ... ***246***, 262
Microchera ... 250, 267
— *albocoronata* ... ***56***, ***84***, ***132***, ***230***, ***250***, 267
Microstilbon ... 254, 272
— *burmeisteri* ... ***254***, 272
Mountain-gem, Green-breasted ... 271
—, **Green-throated** ... ***133***, 271
—, **Grey-tailed** ... ***83***, 271
—, **Purple-throated** ... ***133***, 197, 271
—, **White-bellied** ... ***13***, 271
—, **White-throated** ... ***50***, ***253***, 271
Mountaineer, Bearded ... 263
—, **Eastern** ... ***141***, ***246***, 262
—, Western ... 141, 262
Myrmia ... 254, 272
— *micrura* ... ***157***, ***254***, 272
Myrtis ... 253, 272
— *fanny* ... ***253***, 272

N

Nesophlox ... 255, 273
— *evelynae* ... ***255***, 273
— *lyrura* ... 273

O

Ocreatus ... 248, 264
— *addae* ... 264
— *peruanus* ... 264
— *underwoodii* ... ***19***, *20*, ***137***, ***152***, ***214***, ***248***, 264
Opisthoprora ... 245, 261
— *euryptera* ... ***245***, 261
Oreonympha ... 246, 262
— *albolimbata* ... 262
— *nobilis* ... ***141***, ***246***, 262
Oreotrochilus ... 245, 261
— *adela* ... ***205***, 261
— *chimborazo* ... ***55***, ***218***, ***245***, 261
— *cyanolaemus* ... ***4***, ***207***, 261
— *estella* ... ***108***, 261
— *leucopleurus* ... ***158***, 261
— *melanogaster* ... ***207***, 261
— *stolzmanni* ... 261
Orthorhyncus ... 249, 266
— *cristatus* ... 39, ***95***, ***127***, ***249***, 266
Oxypogon ... 246, 261
— *cyanolaemus* ... 261
— *guerinii* ... ***140***, ***246***, 261
— *lindenii* ... 261
— *stuebelii* ... ***224***, 261

P

Pampa curvipennis ... 267
— *excellens* ... 267
— *pampa* ... 267
Panterpe ... 252, 271
— *insignis* ... ***1***, ***67***, ***133***, ***252***, 271
Patagona ... 249, 265
— *gigas* ... ***13***, *16*, ***177***, ***232***, ***249***, 265
Patagoninae ... 12, 13, 240, 249, 265
Phaethornis ... 242, 256
— *aethopygus* ... 256
— *anthophilus* ... 257
— *atrimentalis* ... 256
— *augusti* ... 257, 274
— *baroni* ... ***9***, 257
— *bourcieri* ... 257
— *eurynome* ... ***242***, 257, 274
— *griseogularis* ... ***13***, 257, 274
— *guy* ... ***74***, ***221***, 257
— *hispidus* ... 257
— *idaliae* ... ***172***, 256
— *koepckeae* ... ***155***, 257
— *longirostris* ... ***33***, ***131***, 257
— *longuemareus* ... ***163***, 256
— *major* ... 257
— *malaris* ... ***89***, *90*, ***102***, 257
— *mexicanus* ... 257
— *nattereri* ... ***274***, 256
— *philippii* ... 257
— *porcullae* ... 257
— *pretrei* ... ***257***, 257
— *ruber* ... ***41***, ***89***, ***103***, 257
— *rupurumii* ... ***89***, 256
— *squalidus* ... ***172***, 256
— *striigularis* ... 257
— *stuarti* ... 257
— *subochraceus* ... 257
— *superciliosus* ... ***60***, 257
— *syrmatophorus* ... 257
— *yaruqui* ... ***155***, 257
Phaethornithinae ... 12, 13, 82, 240, 242, 256
Phlogophilus ... 244, 260
— *harterti* ... 260
— *hemileucurus* ... ***202***, ***244***, 260
Piedtail, Ecuadorian ... 199, 201, ***202***, 203, ***244***, 260
—, Peruvian ... 203, 260
Plovercrest ... 267
—, **Green-crowned** ... ***168***, ***233***, ***250***, 90, 266
—, Violet-crowned ... 90, 233, 266
Plumeleteer, Bronze-tailed ... ***267***, 267
—, **White-vented** ... ***250***, 267
Polyonymus ... 245, 261
— *caroli* ... ***245***, 261
Polytminae ... 12, 13, 240, 243–244, 258
Polytmus ... 243, 258
— *guainumbi* ... ***243***, 258
Polytmus milleri ... 258
— *theresiae* ... ***111***, ***163***, 258
Pterophanes ... 247, 264
— *cyanopterus* ... ***26***, ***143***, ***247***, 264
Puffleg, Black-breasted ... 5, ***201***, 202, ***231***, 262, 276
—, **Black-thighed** ... 199, 201, ***262***, 262
—, Blue-capped ... 263
—, Buff-thighed ... 262
—, Colorful ... 199, 263
—, **Coppery-bellied** ... ***52***, 198, 199, 262, 274, 275, 276
—, Coppery-naped ... 262, 263
—, **Emerald-bellied** ... ***50***, 263
—, **Glowing** ... ***76***, 262, 274, 276
—, **Golden-breasted** ... ***52***, ***246***, 262
—, Gorgeted ... 199, 262
—, **Greenish** ... ***246***, 262
—, **Hoary** ... ***148***, 199, 201, 262
—, **Sapphire-vented** ... ***219***, 262, 263, 276
—, Turquoise-throated ... 201, 262

R

Racket-tail, Booted ... ***19***, *20*, ***137***, ***152***, ***214***, ***248***, 264
—, Peruvian ... 264
—, Rufous-booted ... 264
—, White-booted ... 264
Ramphodon ... 242, 256
— *naevius* ... ***45***, ***173***, ***242***, 256
Ramphomicron ... 246, 261
— *dorsale* ... ***71***, ***138***, ***194***, 261
— *microrhynchum* ... 39, ***41***, ***246***, 261, 274, 275, 276
Rhodopis ... 253, 272
— *vesper* ... ***109***, ***157***, ***253***, 272
Ruby, **Brazilian** ... ***23***, ***30***, ***50***, ***170***, ***248***, 265

S

Sabrewing, Buff-breasted ... 267
—, Curve-winged ... 267
—, Diamantina ... 205, 239, 266, 267
—, **Dry-forest** ... 205, ***239***, 266, 267
—, Dusky ... 164
—, **Grey-breasted** ... 90, ***165***, 239, 266, 267
—, **Lazuline** ... ***250***, 267
—, Long-tailed ... 267
—, **Napo** ... ***22***, 201, 203, 267
—, Outcrop ... 239, 267
—, **Rufous** ... 90, ***266***, 266
—, Rufous-breasted ... 266
—, Santa, Marta ... 138, 199, 200, 267
—, **Scaly-breasted** ... 90, ***266***, 266
—, **Violet** ... 90, ***129***, ***221***, 267
—, Wedge-tailed ... 90, 266, 267
—, **White-tailed** ... 90, ***163***, 198, 267

Sapphire, Golden-tailed *9*, 90, *155*, 270, 276
—, **White-chinned** *29*, 270, 276
Sapphirewing, Great *26*, *143*, *247*, 264
Sappho 245, 260
— *sparganurus* 20, *48*, *159*, *245*, 260
Saucerottia hoffmanni 269
— *viridigaster* 269
Schistes 243, 258
— *albogularis* *39*, *137*, *243*, 258
— *geoffroyi* *137*, 258
Selasphorus 255, 273
— *ardens* 273
— *calliope* *119*, *272*, 273
— *flammula* *273*, 273
— *platycercus* *118*, 273
— *rufus* *119*, *175*, *195*, 273
— *sasin* *17*, *119*, 273
— *scintilla* *255*, 273
Sephanoides 244, 259
— *fernandensis* *204*, 259
— *sephaniodes* *158*, *244*, 259
Sheartail, Mexican *133*, 195, 273
—, **Peruvian** 92, *157*, 204, *253*, 272
—, **Slender** *254*, 273
Sicklebill, Buff-tailed 39, 234, 256
—, **White-tipped** *40*, 90, *235*, *242*, 256
Snowcap *56*, *84*, 90, *132*, *230*, *250*, 267
Spatuletail, Marvelous 20, *27*, 91, *92*, 185, 203, *208*, *223*, *247*, 263
Starfrontlet, Apurimac 263
—, **Blue-throated** *54*, *113*, *213*, 264
—, Bolivian 263
—, **Buff-winged** *27*, *64*, 264, 276
—, **Cuzco** *263*, 263
—, **Glittering** 56, 199, *200*, 264
—, Golden 264
—, **Golden-bellied** *9*, *13*, *113*, 264
—, Huanuco 263
—, Perija 198, 199, 264
—, **Rainbow** *54*, *217*, 263
—, Violet-throated 263
—, **White-tailed** 78, *79*, *99*, 199, 264
Starthroat, Blue-tufted *73*, 271
—, **Long-billed** *253*, *271*, 271
—, Plain-capped 117, 121, 271
—, **Stripe-breasted** *166*, 183, *227*, 271
Stephanoxis 250, 266
— *lalandi* *168*, *233*, *250*, 266, 267
— *loddigesii* 266
Sternoclyta 252, 271
— *cyanopectus* *252*, 271
Streamertail 270
—, **Black-billed** 122, *125*, *226*, 270, *271*
—, **Red-billed** 20, 122, *125*, *251*, 270
Sunangel, Amethyst-throated *50*, *244*, 259, 274, 275
Sunangel, Bogotá 112, 275
—, **Gorgeted** *74*, 259
—, **Little** *259*, 259
—, Longuemare's 259, 274
—, Merida 259
—, **Orange-throated** *64*, 259
—, **Purple-throated** *54*, *217*, 259
—, **Royal** *74*, 78, 201, *203*, 259
—, **Tourmaline** *42*, *105*, *152*, 259
Sunbeam, Black-hooded *224*, 263
—, Purple-backed 203, 263
—, **Shining** *43*, *49*, *56*, *105*, *140*, 263
—, **White-tufted** 203, *247*, 263
Sungem, Horned *13*, *110*, *167*, *243*, 258
Sylph, Long-tailed *18*, 20, *61*, 78, 112, *113*, *228*, *245*, 260, 275, 276
—, Venezuelan 198, 260
—, **Violet-tailed** *86*, *147*, 260

T
Taphrolesbia 245, 260
— *griseiventris* *203*, *245*, 260
Taphrospilus 251, 268
— *hypostictus* *251*, 268
Thalurania 251, 268
— *colombica* *26*, *31*, *71*, *77*, *212*, 268
— *furcata* *160*, 268, 275, 276
— *glaucopis* *171*, *251*, 268
— *ridgwayi* 268
— *watertonii* *206*, 268
Thaumastura 253, 272
— *cora* *157*, *253*, 272
Thornbill, Black-backed *71*, *138*, *194*, 199, 261
—, **Blue-mantled** *261*, 261
—, Bronze-tailed 261
—, **Olivaceous** *140*, 261
—, **Purple-backed** 39, *41*, *246*, 261, 274, 275, 276
—, **Rainbow-bearded** *81*, *101*, 261, 275
—, **Rufous-capped** *246*, 261
Thorntail, Black-bellied 260
—, Coppery 205, 260
—, **Green** *47*, *149*, *244*, 260
—, **Wire-crested** *93*, 199, 201, *202*, 203, 260
Threnetes 106, 242, 256
— *leucurus* *165*, 256, 274
— *niger* 256
— *ruckeri* *242*, 256
Tilmatura 254, 273
— *dupontii* *254*, 273
Topaz, Crimson 20, *72*, 78, 90, 162, 184, *225*, *242*, 256
—, Fiery 78, 256
Topaza 242, 256
— *pella* 20, *72*, 184, *225*, *242*, 256
Topaza pyra 256
Trainbearer, Black-tailed *14*, 93, 261, 275, 276
—, **Green-tailed** *246*, 261, 275
Trochilinae 12, 13, 240, 249–255, 265
Trochilus 251, 270
— *polytmus* 20, *125*, *251*, 270
— *scitulus* *125*, *226*, *271*, 270

U
Urochroa 248, 264
— *bougueri* *146*, 264
— *leucura* *248*, 264
Urosticte 248, 264
— *benjamini* *150*, 264
— *ruficrissa* *248*, 264

V
Velvetbreast, Mountain *247*, 264, 276
Violet-ear, Brown 44, 90, *95*, *146*, *258*, 258
—, **Green** *67*, *75*, 90, 117, 121, 258, 259
—, Lesser 259
—, Mexican 259
—, **Sparkling** *99*, *111*, *141*, 174, 258, 274
—, **White-vented** *243*, 258
Violetear, Lesser 259
—, Mexican 259
Visorbearer, Hooded *96*, 166, *167*, 205, 258
—, **Hyacinth** 166, *167*, *243*, 258

W
Whitetip, Purple-bibbed *150*, 264
—, **Rufous-vented** *248*, 264
Woodnymph, Crowned *26*, *31*, *71*, *77*, *212*, 268
—, **Fork-tailed** *160*, 268, 275, 276
—, **Long-tailed** 168, 205, *206*, 268
—, Mexican 195, 196, 268
—, **Violet-capped** 168, *171*, *251*, 268
Woodstar, Amethyst *100*, *104*, *161*, 272, 275, 276
—, **Chilean** *157*, *204*, *253*, 272
—, **Esmeraldas** *201*, 202, 272
—, **Gorgeted** *254*, *272*, 272, 276
—, Little 201, 203, 272
—, Magenta-throated 272
—, **Purple-collared** *253*, 272
—, **Purple-throated** *74*, *254*, 272
—, Rufous-shafted 272
—, Santa Marta 138, 272
—, **Short-tailed** *157*, *254*, 272
—, **Slender-tailed** *254*, 272
—, **Sparkling-tailed** *254*, 273
—, **White-bellied** *13*, *22*, *150*, 272, 276